Quarterly Essay

CONTENTS

Quarterly Essay is published four times a year
by Black Inc., an imprint of Schwartz Publishing
Pty Ltd. Publisher: Morry Schwartz.

ISBN 9781863954129 ISSN 1832-0953

Subscriptions – 1 year (4 issues): $49 within
Australia incl. GST. Outside Australia $79
2 years (8 issues): $95 within Australia incl.
GST. Outside Australia $155.
Payment may be made by Mastercard, Visa or
Bankcard, or by cheque made out to Schwartz
Publishing. Payment includes postage and
handling.

To subscribe, fill out and post the subscription
card, or subscribe online at:

www.quarterlyessay.com

Correspondence and subscriptions should be
addressed to the Editor at:

Black Inc.
Level 5, 289 Flinders Lane
Melbourne VIC 3000 Australia
Phone: 61 3 9654 2000
Fax: 61 3 9654 2290
Email:
quarterlyessay@blackincbooks.com (editorial)
subscribe@blackincbooks.com (subscriptions)

Editor: Chris Feik
Management: Sophy Williams
Production Co-ordinator: Caitlin Yates
Publicity: Anna Lensky
Design: Guy Mirabella
Printer: Griffin Press

Climate Change
and the Nuclear Option

Ian Lowe

Two years ago, I addressed the National Press Club to explain why nuclear power was not a sensible response to climate change. At the time, some people asked why I was bothering. The Howard government was still in deep denial about global warming, and the nuclear option seemed as dead as a dodo. I gave the speech because I believed it would eventually become impossible for the national government to ignore the reality of climate change. I also feared that it would be consistent with John Howard's approach to other problems for him to devise a noisy distraction – like canvassing the option of nuclear power. My fears proved alarmingly accurate.

As a young scientist, I was enthusiastic about nuclear power. Growing up in New South Wales, I heard regularly of accidents in coalmines. In the 1960s, nuclear power seemed to be a technically advanced means of generating electricity. At the height of pro-nuclear enthusiasm, we were promised that it would deliver electricity so cheaply that power authorities would not bother to meter it! The industry made cheerful promotional documentaries to get us all used to the idea that nuclear power was the way of the future. In 1968, I went to the UK to do research for a doctorate; my project was funded by the UK Atomic Energy Authority through the nuclear scientists designing the prototype fast-breeder reactor at Dounreay. Recognising that uranium is a limited resource, they set out to develop reactors that would produce more nuclear fuel than they consumed. The technical problems proved severe. The US and the UK have since abandoned the whole idea. Only the French effort limps on, more in hope than realism.

When I started lecturing in the faculty of technology at the UK Open University in 1971, I was still a believer in nuclear power, but my views were shaken by some colleagues who asked awkward questions about the economics and about waste management. Back in Australia at the beginning of 1977, I read the Fox Report. Justice Russell Fox, Sir Charles Kerr and Dr Graham Kelleher had been appointed to inquire into the proposed Ranger uranium mine. Their study broadened into a review of Australia's role in the nuclear industry. They wrote:

> In considering the evidence, we have found that many wild and exaggerated statements are made about the risks and dangers of nuclear energy production by those opposed to it. What has surprised us more is a lack of objectivity in not a few of those in favour of it, including distinguished scientists. It seems that the subject is one very apt to arouse strong emotions, both in opponents and proponents. There is abundant evidence before us to show that scientists, engineers and administrators involved in the business of producing nuclear energy have at times painted excessively optimistic pictures of the safety and performance, projected or past, of various aspects of nuclear production. There are not a few scientists, including distinguished nuclear scientists, who are flatly opposed to the further development of nuclear energy, and who present facts and views opposed to those of others of equal eminence.

This is an important point to bear in mind as you read this essay. There is no objective truth about the future performance, cost and safety of nuclear reactors. There is a range of defensible opinions, as well as some that appear indefensible. Even when dealing with the history, some people are selective in choosing evidence that seems to support their position. We are all influenced by our experience, our culture and our values in trying to make sense of complex and uncertain issues. So you should read all statements about the nuclear issue – including this essay – with a critical eye.

The Fox Report made the telling point that nuclear power, while it had been relatively safe and clean until that time as a means of generating electricity, had two fundamental problems: it produced radioactive waste that would need to be stored for immensely long periods, and it provided fissile material that could be diverted to produce weapons. The report argued that it would be irresponsible to contribute to a worsening of these problems without convincing evidence that they had been solved, or were at least likely to be. After considering these arguments, I accepted that I had been wrong to support nuclear power and became more critical.

I now found that the claims about the economic case for nuclear power were very dubious, usually based on careful selection of the past evidence or heroic assumptions about future costs. Back in the UK, I was involved in the late '70s debate about a bizarre proposal by the electricity authority for a crash program to build thirty-six nuclear reactors in fifteen years to avert the coming energy crisis. There was at the time no evidence that an energy crisis was imminent, but when we analysed the demand for concrete, steel and other materials that would be produced by the proposal, we found that it would itself have created a crisis, which the authority would then claim to be solving! So by the time I returned to a permanent appointment at Griffith University in 1980, I had become very jaundiced about the claims of the nuclear power industry.

By then it was clear that nuclear power was expensive, but the industry still had a reasonable safety record and could justifiably claim that it killed and injured fewer workers than did the production of coal-fired electricity. Even this argument was subsequently weakened in 1979 by the Three Mile Island accident; the reactor almost melted down and was effectively destroyed. While good management of the crisis averted a major radiation leak, it is sobering to reflect that the same basic design is used in most of the world's reactors. We were not so lucky with the 1986 Chernobyl disaster, which spread a swathe of radioactive pollution across Europe from the Ukraine to the western parts of the British Isles. That marked the end of public support for the European nuclear power program. The level of nuclear power then steadily declined, as old reactors were retired and not replaced. The Thatcher government tried to prop up the nuclear industry by enacting an obligation for a minimum percentage of power to come from sources other than fossil fuels, but instead this kick-started the UK wind energy industry. By the end of the twentieth century, nuclear power looked like a dying industry.

Then something very strange happened. A small group in the UK nuclear industry concocted the idea of re-badging it as the answer to global climate change. This struck me as a very improbable line. The nuclear power industry had previously used every trick in the book to disparage environmental activists, who had been critical of the industry's record. But desperate times call for desperate measures. The nuclear lobby embraced the science of global climate change, aligning themselves with their old foes such as WWF and Greenpeace. The industry embarked on a very clever campaign of briefing journalists and opinion-makers with the new line: global climate change is a serious problem, clean energy is needed, renewables are unreliable so the world needs nuclear power, which they re-defined as being "clean." Though the claim to cleanliness was dubious, it was seized on by some politicians and journalists. Their enthusiasm was perhaps a sign of desperation, born of a desire either to cling to the old idea of centralised electricity or to find a "silver bullet" for climate change now that the urgency of the issue was plain. This campaign had not yet reached Australia when I spoke to the Press Club, but I was concerned that it might be transferred here from Western Europe. Not long afterwards, the tide turned on public perceptions of global warming and the studied inaction of the Howard government was finally shown by its own polling to be indefensible. Then the Prime Minister returned from Washington in mid-2006 to announce that Australia needed to consider nuclear energy as an option. Interviewed on *AM*, Howard said:

> What I am saying to the Australian people is: let us calmly and sensibly examine what our options are. Let's not set our faces against examining all of those options and when all the facts are in, we can then make judgments. But I don't think all the facts are in in relation to nuclear, because we've had very little debate on this issue over the last twenty-five or thirty years, because everybody's said, oh well, you can't possibly even think about it. That's changed a lot.

It wasn't clear at that point that things *had* changed a lot, but the Prime Minister set about ensuring that they did. A taskforce described by John Clarke as "people who want nuclear power by Tuesday" was hastily put together. The process was so rushed that Howard was only able to give the waiting press the names of some members of the taskforce on the day he announced its formation. In a reminder of the truism expressed by an anonymous American as "Facts ain't given, they're gotten!", the taskforce seems to have set about finding facts that would show the nuclear industry in the best possible light. The subsequent report by Dr Ziggy Switkowski and his colleagues is analysed in more detail below. It was hailed by the Prime Minister and his media cheer-squad as giving the green light for the nuclear industry: "a glowing future" was the Freudian slip in a headline used by the *Australian*. That section of the press even rang me to ask if I had been persuaded by the "rational argument" of the report to "move beyond my emotional opposition to nuclear power." I told them that my opposition to nuclear power was rational and based on both the experience of the last fifty years and a sober assessment of global futures. In this essay I will develop that argument.

Energy is essential for civilised living, but the current approach of basing our energy-intensive lifestyle on fossil fuels is unsustainable. We need to make fundamental changes if our society is to survive. The nuclear option does not make sense on any level: economically, environmentally, politically or socially. It is too costly, too dangerous, too slow and has too small an impact on global warming. That is why most of the developed world is rejecting nuclear power in favour of renewable energy and improved efficiency. We should be a responsible global citizen and set serious targets to reduce our greenhouse pollution, but we should not go down the nuclear path. The rational response to our situation is to combine vastly improved efficiency with an investment in renewable energy technologies.

PEAK OIL AND BEYOND

Energy is the basis of modern civilisation. We have easier lives than our grandparents did because we use much more energy: electricity, gas and transport fuels. The average rate of energy use in OECD countries varies from about 3 to 10 kilowatts for every person. One way of visualising this is to imagine that our average energy use in Australia is equivalent to about forty human slaves working for us in shifts, doing what slaves used to do: produce our food, carry us about, wash our clothes, entertain us, fan us when we are hot, provide hot water for washing and so on. Energy is also used to ease other shortages. Cities with limited drinking water are now using desalination plants to process sea water – using more energy. We have increased food supply for our growing population by farming more intensively – using still more energy. As we exhausted rich metal ores, we moved on to poorer deposits – but that takes much more energy. Without usable energy, modern societies would literally grind to a halt. We see occasional short-term demonstrations of this, as when a bushfire interrupted the electricity supply to the city of Melbourne early this year, causing chaos. It is not easy to imagine how our cities would cope if some natural disaster or extreme political event were to halt the flow of oil to Australia, given the total reliance of our transport system on petroleum fuels.

We now face two serious problems. Experts disagree about whether we are approaching the peak of world oil production, or have actually passed it. Either way, we are near the end of the age of cheap petroleum fuels. The second problem is that the present use of "fossil fuels" – coal, oil and gas – is changing the global climate. Both problems are compounded by huge inequalities. On average, we use about half as much energy as US citizens, but about ten times as much as Chinese and a hundred times as much as people in the poorest parts of the world. This unfair access to the benefits of energy-intensive lifestyle choices is creating tension, especially as the disadvantaged are made ever more aware of the disparity by US television and film. A sustainable world will have to be based on much more equitable access to the energy services we take for granted.

We have known about the problems of peak oil and climate change for decades. Experts told us in the 1970s that world oil production would peak in about 2010. Thirty years ago, I gave public lectures and published papers calling for a national energy policy in general and a strategic approach to the particular problem of transport fuels. Similarly, the science was telling us twenty years ago that climate change is a real threat to civilised society, demanding a new approach to energy supply and use. But few countries have an overall energy policy to guide the transitions from cheap oil and large-scale coal use. As recently as 2006, the Australian government was still effectively in denial about climate change, and it still behaves as if peak oil were an obscure scientific theory rather than an urgent policy imperative. Future generations will find it difficult to understand our collective incapacity to acknowledge or respond to these problems.

Our failure to develop concerted responses to future energy needs is simply one facet of a more general inability to think about the sustainability of our lifestyle choices. In 1999, the UN's Environment Program report on the global environmental outlook said: "The present approach is not sustainable. Doing nothing is no longer an option." A sustainable society would not erode its resource base, not cause serious environmental damage and not produce unacceptable social problems. Our present lifestyle does not satisfy *any* of these criteria. We are dissipating resources future generations will need, damaging environmental systems and reducing social stability by widening the gap between rich and poor. It is possible to move to a sustainable future, but it requires fundamental change in our values and social institutions. While the United Nations Environment Program said that doing nothing about the huge problems we face is not an option, it is fair to say that it remains the most common response of today's decision-makers.

The idea of "peak oil" was developed in the 1950s by King Hubbert, a US geologist who noticed that the graph of oil production over time was very similar to that of oil-field discoveries, with a time lag. That makes intuitive sense: companies discover oil deposits, map the resources, raise capital to establish production wells and then start extracting the oil for sale. Hubbert predicted that oil production from the continental United States would peak in the early 1970s. No decision-makers took his ideas seriously until the peak actually happened in 1971. The organisation of oil-producing countries, OPEC, was then able to drive a harder bargain because the US became a major purchaser, shifting the balance between sellers and buyers. World oil prices increased in eight years from US$2 per barrel to US$30. The economic impact was huge because transport fuels are used in the production and distribution of almost everything we use. The prise rise also had a dramatic effect on energy use generally, because oil is not just used for transport in the northern hemisphere. North America makes substantial use of oil for heating. France and Japan had moved from coal to oil for electricity supply in the 1960s because oil had been cheaper and people were increasingly unwilling to risk their lives mining coal. Those countries suddenly found oil in short supply and very expensive, precipitating a rapid move in both countries to nuclear energy. This is the basic reason why France is the world's most intensively nuclear nation; having converted their electricity supply to oil and closed down their coal industry, they were left with very few options when the OPEC crisis hit and pushed up oil prices to make oil-fired electricity prohibitively expensive.

After the validity of the peak oil theory was proven, analysts started using the same approach to model global production. Thirty years ago, I and others drew attention to the tentative conclusion of this work: world oil production was estimated to peak between 2000 and 2020, with a median estimate of 2010. This gave us almost thirty years to plan a transition, the likely amount of time it would take to manage the complete reorganisation of our transport system. Yet today billions of dollars of public money are still lavished on extravagant and ugly road developments, based on an implicit assumption that the age of cheap fuel will last forever and we will continue driving ourselves in large inefficient cars. There are several technical alternatives to oil. In fact, alternative transport fuels have been known and used for decades, including alcohols from plant material and synthetic liquid fuel from gas. But many of the short-term options, such as oil from shale or coal-to-oil conversion, would dramatically exacerbate the greenhouse pollution from transport fuels. Looking further ahead, we see some clean alternatives, most obviously fuel cells using hydrogen produced by splitting water with electricity from renewable supply technologies, but all these options are sure to be much more expensive. We should be planning now for the post-petroleum age.

Australia is already what Barry Jones called a post-industrial economy. The traditional sectors of agriculture, mining and manufacturing now account for less than a quarter of the economy and fewer than 20 per cent of jobs. Most of us work in the services sector, if this term is applied broadly to cover activities such as advertising. Our relatively high wages and salaries make our production costs high in labour-intensive fields like textiles, clothing and footwear, so these activities have largely been exported to low-income countries; in the last ten years, Australia has shed more than 150,000 manufacturing jobs. Our low level of investment in research and innovation makes us generally uncompetitive in value-added manufactures, apart from a few niche areas. For seven years we have had a trade deficit every month, largely because we export low-value commodities and import value-added goods and services.

The knee-jerk response from economists and politicians is to urge us to invest more in infrastructure to allow greater volumes of low-value commodities to be exported, locking us securely into the economic structure of a Third World country, or the only state worse than Paul Keating's "banana republic": a banana constitutional monarchy. As I was writing this essay, the financial press was vigorously urging state governments to spend more public money to accelerate the export of our low-value commodities such as coal and iron ore. In the short term it is possible to create the illusion of economic progress by liquidating our natural capital, but there is a fundamental economic problem with such an approach. Mining operations are inevitably boom-and-bust because the resources are systematically depleted.

Consider, by contrast, the forward-looking decision-makers in the Canadian province of Alberta. Thirty years ago, when their agrarian economy was dramatically transformed by the discovery of oil and gas resources, the provincial authority decided to invest some of the profits in the Alberta Heritage Trust Fund, which was used to establish future economic opportunities for the day when the oil and gas would be gone. The trustees invested the mineral royalties into a range of employment alternatives, ranging from plantation forestry to micro-electronics. Alberta was looking ahead, recognising that a mineral economy had a limited life and investing in its own future. In Australia, both state and Commonwealth governments have used the short-term bonanza to fund extravagant lifestyle choices or political hand-outs, with little or no effort made to plan for a future in which current resources are exhausted. So we are heading down the Nauru path to a depleted moonscape, rather than taking the Alberta approach. This issue has a particular significance for the nuclear debate. While the late George Seddon correctly argued that the modern mining industry is knowledge-based, increasing Australian involvement in uranium export would still reinforce this broad approach of making us Quarry Australia, paying for our imports by the export of low-value bulk commodities.

Fifty years ago, Australia was one of the most equal societies in the world. Today, we are one of the most unequal of all the industrialised nations. Again, the mineral export industries are influential. The industry is capital-intensive and commodity prices have been driven up by demand in China, so mining companies can afford to pay very high wages compared with manufacturing or services. In cities like Perth, the high incomes of mining workers on fly-in-fly-out contracts are driving house prices to levels that put the Australian dream beyond the reach of more and more young families. While the good times of the mining boom roll on, we should be using the revenues to invest in a cohesive social future, rather than allowing the windfall gains to widen inequality.

We should also be responsible global citizens. It is our humanitarian duty to improve the lot of the poorest people of the world. It is also enlightened self-interest because a world of increasing inequality will be a world of increasing tension. As the Australian delegation said at the 1999 UNESCO World Conference on Science, we should aim to make this not just a new century but a *just* new century. If we are serious about deciding who will come here – Howard's campaign slogan in 2001 – we should also accept that desperate people who can't provide for their families will be prepared to risk their lives in search of a better world. The best way to stop people braving shark-infested waters in leaky boats is to ensure they have a decent life where they now live. Several studies predict large-scale displacement of people in our region by climate change unless we take concerted global action.

Around the world 1.2 billion people still need clean water, about twice that number don't have decent sanitation, 800 million are hungry and hundreds of millions lack proper shelter. To meet their basic needs within the limits of global systems, we must use cleaner energy and use it more efficiently. The resources are there to provide every person with a decent life. A recent study by the United Nations Development Program concluded that everyone could have clean drinking water, adequate nutrition and shelter, basic health care and education, for about 5 per cent of the global military budget. So the problem is clearly political and social as much as technical. The notion that we need nuclear power to provide energy for the developing world runs parallel with the notion that we should introduce genetically modified food crops to feed the hungry; these purported solutions presume that we accept the present inequality of distribution and propose instead an increase in supply. If the sink won't fill because the water is running down the drain, it isn't very clever to keep turning the tap. Putting in the plug is a better course of action.

We should also recognise the link between ecological systems and human health. We rely on natural systems to provide the essentials of life: oxygen, water and food. We also need those systems to process our waste. When the natural world's capacity to provide these essentials is run down, this has direct, systemic effects on our health. More broadly, the natural world provides us with our sense of place, cultural identity and spiritual sustenance. We are healthier and more fulfilled when such needs are satisfied. An investment in the health of our natural systems is also an investment in the health of the human community.

In my book *A Big Fix*, I argued that existing policies will not do because they fail to address the pressures of a growing population and rising material demands. We need to stabilise the population by setting responsible immigration targets and ending the ridiculous incentives to have more children. A sustainable future will involve using resources more efficiently, maintaining natural systems, developing social cohesion, nurturing our cultural traditions and finding durable economic activities. Balancing those dimensions is a complex task that defies simplistic approaches, such as trusting the market or trying to keep everything the way it is today. We have to take difficult decisions about which things must change and which should be preserved. This will only be possible if there is an open and transparent process — one that is democratic and participatory and which allows time to work through the costs and benefits of alternatives. Changing one thing in a complex system always produces other changes, so no change is ever universally beneficial; there are always losers as well as winners. In a fair world, those who lose out from a change that benefits the community as a whole should be compensated. We accepted that principle when we decided that those using Sydney Airport should pay a noise levy to compensate those under the flight path. It is a principle we should apply more generally, identifying the winners and losers from new policies and establishing mechanisms to ensure the winners compensate the losers.

This is a specific example of a more general conclusion: moving toward a sustainable future will require new social and political institutions. Concerted action is frustrated by the division of responsibility between different levels of government, by state and local boundaries that have no social or ecological logic, and by the silos of established departments. We need new structures that will enable co-ordinated policies and actions, integrating environmental goals with our social, economic and cultural aspirations.

THE GREAT GLOBAL WARMING CHALLENGE

On 17 January 2007, the *Bulletin of the Atomic Scientists* moved the hands of its famous doomsday clock forward so they now stand at five minutes to midnight, in recognition of the twin threats of climate change and associated risks of nuclear proliferation.

Was it a little slow off the mark? Twenty years ago it was clear to many that the global climate was changing. In 1989 I wrote a book on the topic, *Living in the Greenhouse*. But at the time the scientific community was still divided; some cautious colleagues were reluctant to accept that the human use of fossil fuels was actually *causing* the changes that were being observed. This is an important distinction, as much dishonest political rhetoric implies a causal link between events that are only related in time: "Since we have been elected ..." is the usual cry from those in office to claim credit for anything good that has happened, whether they had any influence over it or even knew it was occurring, while "Since this government was elected" is used with equal chutzpah by those in opposition to blame the government for everything from droughts, fires and floods to water shortages. As more research has been done, it has become clearer that changes to the global climate have not just followed the increasing human use of fossil fuels but are being *caused* by that process. While there was genuine uncertainty in the science, it was defensible (if short-sighted) for politicians to do nothing about the problem. The science has now been refined to the point where there is no serious dispute about the human influence on climate.

Nevertheless, the idea that there is still uncertainty is being spread by far-right lobby groups like the Institute of Public Affairs and some of their apologists in the media. Recently the ABC was praised by these people for showing a UK-produced documentary called *The Great Global Warming Swindle*. It was a clever mix of some science, many half-truths and some claims that were nonsense. The version originally shown in the UK made the amazing claim that volcanoes release more carbon dioxide than all human fuel use, when the real average figure is about 2 per cent of human production. While that falsehood and some misrepresentations that provoked legal action were removed before the Australian screening, the documentary still contained false assertions. It claimed medieval times were much warmer than today; the science shows some regions were briefly warmer, but average global temperatures were lower. It claimed changes in sun-spot activity explain the recent warming, based on a Danish scientist's analysis of pre-1990 evidence, when more recent data disprove the link. It had Professor John Christy, director of the Earth System Science Center at Alabama, stating that temperature changes in the atmosphere are inconsistent with the theory of global warming, when Christy has since admitted that claim was based on incorrect data. It claimed average global temperatures declined dramatically between 1940 and 1976, suggesting this cooling debunks the science. Our hemisphere actually warmed during that period. Pollution from coal burning caused a slight decrease in the northern hemisphere, so there was a very small overall decline, but not a dramatic cooling that would cast doubt on the science.

Other claims were true but misleading. The Earth has warmed in the past, due to natural variations in our orbit and the sun's radiation, with related increases in carbon dioxide levels. This shows warming can release carbon dioxide from oceans and vegetation. It doesn't refute the science that found in the 1890s that putting extra carbon dioxide into the atmosphere traps more heat. The film's fundamental point was that the climate has varied naturally in the past, so there is no reason to believe we are causing the recent changes. Of course there have been natural variations, but the only models that accurately represent climate change take account of both natural changes and the recent human influence. As I finalised this essay, it was too early to tell whether this piece of disinformation had influenced the public discourse. It won't affect the "debate" because there is no longer any serious argument; the science of climate change is now comprehensive and compelling.

The politicians of the developed world accepted the scientific arguments ten years ago when they negotiated the Kyoto agreement to slow down the release of greenhouse gases. Only the Bush regime in Washington and the Howard government in Canberra have refused to ratify this protocol. The Earth as a whole has warmed about 0.8 degrees Celsius in the last hundred years, with Australia warming slightly more than the global average. The Earth is now warmer than at any time since credible records began to be kept. As predicted by climate scientists, there have been other changes associated with the warming: shrinking of glaciers, thinning of polar ice, rising sea levels, changing rainfall patterns and more frequent extreme events such as droughts and severe storms. Climate change is already having serious economic effects: examples include reduced agricultural production, increased costs of severe events (fires and storms) and the need to consider radical water supply measures such as desalination plants. Of course, climate change doesn't just have short-term economic effects. The Millennium Assessment Report, released in 2005 by the United Nations, warned that species loss is accelerating. In part this is due to the growing pressures of habitat loss, introduced species and chemical pollution, but these forces are now being supplemented by climate change. The report warns that we could lose between 10 and 30 per cent of all mammal, bird and amphibian species this century. These alarming consequences have driven distinguished scientists, including James Lovelock, to conclude that the situation is desperate enough to reconsider our attitude to nuclear power. I agree with Lovelock about the urgency of the situation, but not about that response, which will create a range of new problems.

Just as other serious environmental problems have been tackled at the international level, a concerted global response to climate change is needed. The science shows that the world as a whole must reduce greenhouse gas emissions to about 40 per cent of present levels (or less) by 2050 (or sooner). Cutting our own country's emissions is our obligation to the world community. It will require action from all levels of government to encourage cleaner energy supply and much more efficient conversion of energy into the services we need. The obvious way to fund the transition is systematically to phase out the huge current subsidies of fossil-fuel supply and use, transferring these public funds to the expansion of renewable energy supply technologies and efficiency gains. Various studies estimate the annual public subsidy of fossil-fuel supply and use in Australia as between five and eight billion dollars (without allowing for the costs of climate change).

There are three ways to reduce the amount of carbon dioxide we put into the air. First, we must use cleaner fuels. Gas is far preferable to electricity, especially in those countries where most of the electric power comes from burning coal. Using coal-fired electricity to heat water or cook, rather than burning gas, puts about four times as much carbon dioxide into the air. Renewable energies, such as solar or wind power, release very little carbon dioxide, so they should be the most preferred option.

The second part of the solution involves turning energy more efficiently into the services we want. Nobody actually wants *energy*; we want hot showers and cold drinks, the ability to cook our food, wash our clothes and move around. Most of the technology we use is very wasteful. The European Union now has a target of cutting energy use by a quarter by 2020, and some countries, such as the Netherlands, have more ambitious aims. Efficiency improvements should be a universal goal in OECD countries.

Energy efficiency provides economic benefits because saving energy is much cheaper than buying it. The recently published Australian book *The Natural Advantage of Nations* gives a number of case studies. In the last decade the multinational chemical firm DuPont has cut its energy use by 7 per cent and its greenhouse pollution by over 70 per cent while increasing production almost 30 per cent. It saved more than $2 billion in the process. Five other major firms, including IBM, Alcan, Bayer and British Telecom, have reduced their greenhouse gas emissions by 60 per cent since the early 1990s — and saved another $2 billion. In 2001, BP announced that it had already met its 2010 target of cutting greenhouse gases to 10 per cent below its 1990 level. It reduced its energy bills by $650 million over the decade. General Electric has set a goal of improving energy efficiency by 30 per cent by 2012. At the extreme end of the range, silicon chip company ST Microelectronics has set a target of zero net carbon emissions by 2012. Improving efficiency makes business sense. And at the household level, if your fridge or washing machine is more efficient, that is real money in your pocket as well as a win for the environment.

Finally, we may need to reduce global consumption of goods and material resources, a subject to which I will return later in this essay.

Promoting nuclear power as the solution to climate change is like advo-cating smoking as a cure for obesity. That is, taking up the nuclear option will make it much more difficult to move to the sort of sustainable, eco-logically healthy future that should be our goal. It is axiomatic that we should be heading in that direction and the political impulse to achieve it should be a bipartisan one. Yet, as I have noted, all the important indi-cators are going backwards. The nuclear option would be a further, deci-sive step in the wrong direction. The essential pro-nuclear argument is that massive amounts of energy will be needed to power our future, that fossil fuels are now recognised as too dirty, that none of the clean energy alternatives can fulfil that demand and so nuclear is the least worst option. Here, in condensed form, is why I think that argument is specious.

First, the economics of nuclear power don't stack up. The true cost of nuclear electricity in Australia would certainly exceed that of wind power, energy from bio-wastes and some forms of solar energy. Geothermal energy from hot dry rocks also promises to be less costly than nuclear. That is without including the large and still uncertain costs of decommis-sioning power reactors and storing the radioactive waste.

So there is no economic case for nuclear power. Government ministers recognised this until the Prime Minister's line changed and they felt obliged to dance to a new tune. In his book *High and Dry*, Guy Pearse notes that in 2005, the industry minister, Ian Macfarlane, was saying, "You don't have to be a rocket scientist to work out [that it is] very, very hard for domestic nuclear power to stack up economically," and that in 2006 the finance minister, Nick Minchin, claimed that nuclear power would not be viable in Australia for at least a hundred years! By the end of 2006, they were dutifully parroting the new line, with Minchin saying, "Nuclear power must be seriously contemplated if you are serious about green-house emissions," and Macfarlane going even further out on the same limb by declaring, "We would be silly not to consider the nuclear option [because] only nuclear power has the potential to deliver baseload quantities of cleaner energy in Australia."

This is a delusion. As energy markets have liberalised around the world, investors have turned their backs on nuclear energy. The number of reactors in Western Europe and the US peaked fifteen to twenty years ago and has been declining ever since. By contrast, the amount of wind power and solar energy is increasing rapidly. In the decade up to 2003, the average annual rates of increase of the different forms of electricity supply were as follows: wind increased by almost 30 per cent, solar by more than 20 per cent, gas 2 per cent, oil and coal 1 per cent, nuclear 0.6 per cent. The figures tell the real story. Despite the recent pro-nuclear hype, most of the world has rejected nuclear energy in favour of alternatives that are cheaper, cleaner and more flexible. These figures also refute one of the oft-repeated lies about the Kyoto agreement. As recently as June, Christopher Pearson in the *Australian* was trotting out the tired old line that Kyoto was devised by the cunning Europeans who can easily meet their targets because they use nuclear power. Most European countries have the same amount of nuclear power now as they had in the Kyoto base year, 1990 (the year against which future emissions are measured). Some have less. Finland is the only European country I am aware of that has commissioned a nuclear reactor this century. So any carbon benefit flowing from use of nuclear power was already there in the European baseline and does absolutely nothing to make these countries' targets any easier to reach. In fact, it is easier for Australia to reduce its emissions precisely because so much of our energy now comes from coal-fired electricity. We could produce the same amount of energy while releasing less carbon, just by moving from coal to gas.

Second, nuclear power is far too slow a response to the urgent problem of climate change. Even if there were political agreement today to build nuclear reactors, it would be at least ten years before the first such reactor could deliver electricity, while some have suggested that between fifteen and twenty-five years is a more realistic estimate. We can't afford to wait decades for a response given the heavy social, environmental and economic costs that global warming is already imposing, especially when much more immediate and appropriate responses are already available. If we were to start today expanding the use of solar hot water in Queensland to cover half of the households – a similar level to the Northern Territory – we could save about as much electricity as a nuclear power station would provide, and do it years before any reactor would be up and running.

The third problem is that nuclear power is too dangerous. Not only is there the risk of accidents such as at Chernobyl, there is also an elevated risk of nuclear weapons proliferation or nuclear terrorism. As far back as 1976, the Fox Report warned that exporting uranium would inevitably increase the risk of nuclear weapons being developed. Since then the situation has steadily worsened, with nuclear weapons having been developed by a range of countries. The experience of Iraq, being invaded by the US and its coalition of the willing, is clearly spurring on Iran and North Korea to develop their own nuclear deterrents. Dr Mohamed ElBaradei, Director General of the International Atomic Energy Agency (IAEA), told the 2005 review conference on the Nuclear Non-Proliferation Treaty (NPT) that "fears of a deadly nuclear detonation … have been re-awakened." He went on to warn of "vulnerabilities in the NPT regime" and the limitations in the authority's power to verify compliance. Despite Dr ElBaradei's passionate pleas, for which his agency was awarded the Nobel Peace Prize, the UN conference ended in utter disarray. The chairperson was not able even to produce a final statement summarising the areas of disagreement. Most of the states holding weapons and some others aspiring to join the nuclear "club" are clearly in breach of the treaty. China and the US are developing and testing new weapons. The existence of nuclear weapons or programs aimed at their production lends an extra dimension of instability to the international "hot spots" of the Middle East, the Korean peninsula and the Taiwan Strait.

The growing threat of terrorism makes the problem even more acute. The willingness of desperate people to engage in acts of gratuitous violence makes it imperative to protect all elements of the nuclear fuel cycle, including waste management, in military fashion. This inevitably adds to both the economic costs of nuclear power and the social costs of using the technology. Embracing the nuclear fuel cycle would both increase insecurity and justify further erosion of our civil liberties, which have already been systematically curtailed on security grounds.

The danger also takes another form. Nuclear power necessarily produces radioactive waste that has to be stored safely for hundreds of thousands of years. After nearly fifty years of nuclear power, the world has produced more than 250 million tonnes of radioactive waste, with some 10,000 tonnes of it highly radioactive, yet nobody has found a permanent solution to the storage problem. In the absence of such a solution, expanding the rate of waste production is irresponsible. The Fox Report suggested thirty years ago that any large-scale use of uranium should await an adequate response to this problem. It is important to emphasise that it is not solely a technical matter of developing systems that will isolate high-level waste for over 200,000 years. The storage issue is also a huge challenge for our social institutions, as it obliges us to consider a time-scale much longer than any human society has lasted, of the same order of magnitude as our entire existence as a species. As AMP Capital Investors said in their 2004 *Nuclear Fuel Cycle Position Paper*:

> there are significant concerns about whether an acceptable waste disposal solution exists ... [W]hile the nuclear waste issues remain unresolved, the uranium/nuclear power industry is transferring the risks, costs and responsibility to future generations.

Fourth, nuclear power is not carbon-free. Significant amounts of fossil-fuel energy are used to mine and process uranium ores, enrich the fuel and build nuclear power stations. Over their operating lifetime, nuclear power stations release much less carbon dioxide than does the burning of coal, but in the short term they would make the situation worse; building nuclear power stations would actually increase greenhouse pollution.

A fifth, and related, problem is that high-grade uranium ores are limited. On best estimates, known high-grade ores could supply present demand for about fifty years. If we expanded the nuclear contribution to global electricity supply from the present level – about 15 per cent – to replace all coal-fired power stations, the high-grade resources would only last for about a decade. There are large deposits of lower-grade ores, but these require much more conventional energy for extraction and processing. Total life-cycle analysis has concluded that fuelling nuclear power stations from lower-grade ores actually releases more carbon dioxide per unit of delivered energy than burning gas! These calculations are disputed by pro-nuclear activists, but there is no doubt that the fuel energy, consequent greenhouse emissions and the dollars needed to produce uranium all increase rapidly as the ore grade declines. If nuclear power were to expand on the scale demanded by some of its proponents, the energy demand and the associated pollution would be considerable.

DOES THE REST OF THE WORLD NEED IT?

Even if we don't need nuclear energy, it is sometimes claimed that the rest of the world is not in our happy position. The 2006 report of the House of Representatives Committee on Industry and Resources, populated by Liberals and some pro-nuclear elements of the ALP, was tendentiously titled *Australia's uranium – Greenhouse friendly fuel for an energy hungry world*. The implication was clear: small children will freeze to death in the dark in less fortunate countries if we deny them the wonderful alternative of uranium. This argument is usually based on an extrapolation of existing energy use. An extreme example of it can be found in a short book by the British scientist and science fiction writer Fred Hoyle. In *Energy or Extinction* (1977), he claimed that the standard of living in the US was higher than in the UK because of the much greater level of energy use. If the British could increase their energy use to the US level, he said, the standard of living would inevitably rise to match that enjoyed by Americans. The argument is false. Unkind critics said Hoyle should have stuck to science fiction, while a more trenchant one said he had! If, to take an extreme example, it were possible to wave a magic wand and cause every vehicle on the road to use twice as much fuel as it now does, our transport energy use would have doubled but few people would feel their standard of living had improved. If your refrigerator used twice as much energy as it does now, or you replaced efficient lights with the old incandescent bulbs, you would increase your domestic electricity use but you would not feel your standard of living had risen proportionately. In fact, because you would be paying more for the same service, you would probably feel that your standard of living had actually declined. The fundamental reason why the level of energy use is much higher in the US than any other country is that the US is the most wasteful nation on Earth. There is a link between energy services and material wellbeing, but the link with *total* energy use is much weaker.

Hoyle assumed that in a future world of ten billion people, each person would want to use energy at the US rate. He concluded, unsurprisingly, that fossil-fuel resources could not sustain such an inflated rate of energy use for very long. Therefore, he argued, the only answer is large-scale adoption of nuclear power. A similar, if less crude, argument has been used by the World Energy Council, an umbrella group for the energy industries, and authors such as Gwyneth Cravens, who incorporate the new reality of climate change: the world needs much more energy, we can't expand fossil-fuel use, so it has to be nuclear power. "If we don't make the right choices, if energy becomes scarce or really expensive, people will give up their big houses and their lawns," Cravens quoted a pro-nuclear activist as saying; but this disastrous situation can be avoided if we embrace nuclear energy. In fact, the World Energy Council's answer to climate change is to allow greenhouse pollution to keep increasing until 2030 and try to return it to present levels by 2050, a hopelessly inadequate response that would condemn the world to very dangerous temperature increases. On that basis it has argued for continuing to use more coal, calling the approach "pragmatic and realistic." Their report argued that wind and other forms of renewable energy "could not be deployed quickly enough" to reduce the carbon dioxide associated with electricity generation, so "we need to build nuclear." The Council's report poured cold water on the other technocratic delusion of carbon capture and storage from coal-fired power stations, pointing out that it would require "capturing and moving a quantity of liquefied gas, the volume of which is equivalent to all the oil and gas movement in the world today, and we have no infrastructure for it." It noted that the test of public acceptance for geological storage is "in much the same category as nuclear waste." On the other hand, the Council saw no problems, either technical or in terms of public acceptance, in large-scale use of nuclear power. It also seems to think, remarkably, that nuclear power is a rapid alternative that is needed because wind turbines and solar energy devices "could not be deployed quickly enough." This borders on the dishonest, given the likely time-scale for building nuclear power stations is a decade or more, while wind turbines can be built in less than a year and a solar hot-water system could be on your roof next week. This is a fundamental flaw in the argument that nuclear power could be part of the solution to global climate change: it is not feasible to build reactors fast enough.

The underlying problem with the argument is the assumption that in a future world of nine or ten billion people, all will live as wastefully as Americans now do. It should not be necessary to point out that this is physically impossible. US resource use would, if extrapolated to the present global population, require about ten planets. There is no prospect even in principle of that sort of extravagant lifestyle becoming universal. If you assume an unrealistic level of global energy use, you are more likely to conclude that conventional resources are inadequate and accept the desperate strategy of building nuclear power stations. But the scale of that task would be equally impossible. Hoyle's future world energy use, which he envisaged as being supplied by nuclear power, would have required the building *every day* of a new large power station – and the decommissioning every day of one earlier installation that had reached the end of its useful life. The scale of uranium mining, fuel fabrication and waste management involved would be absurdly unrealisable, especially if we take into account the limited resources of high-grade uranium ores. The mine at Roxby Downs, in the north of South Australia, supplies a small fraction of the world's present uranium demand. The scale of uranium use envisaged by the nuclear optimists would require literally hundreds of mines of similar size – all using fossil-fuel energy and all producing mountains of low-level waste in the form of radioactive mine tailings.

The fact that quite intelligent people embrace nonsense of this kind reveals a deep-seated myth, based on an underlying set of values that constitute the real problem for human civilisation. If you believe that our material well-being is related to our heavy resource use, and realise that there are literally billions of people in the world living with much less, you are drawn to the case for a huge increase in resource use to allow them to live as we do. If you then do the sums and calculate the resources needed for that expansion, using present technology, you conclude that it is not possible. So you must either decide that we have to live more simply so that others may simply live – as I believe to be the rational conclusion – or postulate some sort of technical fix. Those who are drawn to the hope of a technical fix are intrinsically unlikely to trust solar, wind or other renewable energy supply options, because they seem less likely to be under centralised administrative control. So they form an emotional attachment to nuclear power, or carbon capture and storage, or both, seeing these as technical solutions to the problem. Once having made that commitment, they fall prey to the litany which justifies their chosen answer: renewables can't be scaled up to meet our needs, they can't be deployed quickly enough, their construction uses more energy than they would produce, and so on. Regardless of how often these claims are shown to be urban myths, they keep being repeated because they satisfy a deep-seated belief system. I think I understand the psychology behind those beliefs, but they are no substitute for a rational energy policy.

It is neither possible nor desirable for all humans to live as wastefully as Americans now do. It is possible for all people to live at the level of resource use that prevailed in Australia in the 1960s; not a time of Neolithic privation, but a less wasteful era than the present one. We lived in smaller dwellings, each on average occupied by more people, we used less electricity and water, we were much more likely to use public transport or small efficient cars, we ate more fresh produce and less processed food. It is entirely possible for a mix of renewable energy supply systems to provide enough energy, if turned efficiently into services, to allow all of the Earth's people to live comfortably. Renewables now provide a third of Sweden's energy, half of Norway's and three-quarters in the case of Iceland, which has decided to be totally powered by renewables by 2020. While there is some scope for grid systems to be deployed in countries with large populations in a contiguous landmass, like China or India, there are many developing countries for which either the physical geography or the population distribution dictates a different approach. Indonesia is a country of more than 15,000 islands, while the Philippines has about 7000. If the people in those nations are to have the sort of energy services we take for granted, they will have to be based on local renewable energy supply.

We are often urged to consider the impact of rapid industrialisation in China on the world's climate. China is building nuclear power stations, but it is also investing massive amounts in renewables, especially wind and solar, planning to get more than twice as much energy from these technologies as from nuclear. More importantly, the Chinese leadership understands the fundamental principle that a sustainable future involves real changes. At the 2005 conference, "Sustainable Development – China and the World," I heard the Chinese leaders expound the principle of the "three zeroes": zero growth in population, zero growth in resource use and zero growth in pollution. Like Western leaders, they could well be accused of having a gulf between their stated aspirations and the results of their day-to-day policy choices. They concede that it will be no small achievement to match those goals to the material aspirations of their country's people, but the goals contrast dramatically with the naive emphasis on perpetual growth in our political culture.

I acknowledge that there are about 440 nuclear power stations in operation around the world. Almost all of them are more than twenty-five years old, so many are now approaching the end of their design lifespan. Pro-nuclear advocates are correct to say that more carbon dioxide would now be going into the atmosphere if all of these stations had been coal-fired instead. Of course, the atmosphere would have been spared the same amount of carbon dioxide if we had an equivalent mix of renewable energy supply systems, or if the demand for electricity were reduced by an equivalent contribution from solar hot water. Appropriate comparison is critical. It is sometimes claimed that exporting more Australian uranium will slow the rate of global climate change. It would have that short-term benefit if the only result were to cut the number of coal-fired power stations. On the other hand, I believe that it is now clear to any responsible decision-maker that we should not be worsening the problem by burning coal. So the real choice is between nuclear power and a mix of renewable energy technologies combined with efficiency measures. If that is the choice, it would take creative arithmetic to make a case that our uranium is doing anything at all to save the world from climate change. I would be more impressed by the integrity of those arguing for us to export uranium to slow down global warming if they were also calling for us to reduce our coal exports. Australia could do much more to help the global atmosphere by cutting our coal exports than we could by the most fanciful estimate of the potential benefits from our uranium. Of course, many of those urging uranium exports are also in the vanguard of calls to export even more coal than we do today. This shows that they are actually more interested in the short-term economic benefits of mineral exports than in any effect on the global environment.

As I was writing this essay, the state government of Queensland was actually *celebrating* a plan for a massive increase in the scale of coal exports, while the New South Wales government also seemed keen to expand its contribution to the global problem. It is hard to think of a more striking example of governments pursuing short-term economic goals at the expense of our future. Exports of both coal and uranium from Australia are dominated by the same two giant mining companies, Rio Tinto and BHP Billiton. Neither shows much interest in scaling back their operations for the good of the global environment.

A few technocrats have steadfastly advocated that Australia should mine and export uranium, enrich it for export, build nuclear power stations and (in some cases) even offer to take in the world's radioactive waste. These people had their day in the sun when John Howard assembled his taskforce in June 2006.

While the Switkowski report, *Uranium Mining, Processing and Nuclear Energy – Opportunities for Australia?*, was hailed by some as a green light for the nuclear industry, a detailed reading shows that it is a very lame endorsement of the nuclear option. The whole exercise was of doubtful merit. The taskforce was hand-picked by the Prime Minister, certainly giving the appearance that the members had been chosen to give the result he wanted, while the terms of reference limited the study by excluding consideration of emerging renewable energy alternatives and limiting environmental considerations to the possible reduction of greenhouse gas emissions. But the facts, even if carefully chosen, still speak for themselves. Despite the brave show made by John Howard on releasing the report and some of the media commentary, which was clearly based on the government briefing rather than a careful reading of the report, the fine print shows that nuclear power is an expensive, slow and dirty way of making very little impact on the problem of global climate change.

First, consider the economics of nuclear power. It is hard to make accurate comparisons between different forms of electricity generation because the sums depend critically on the underlying assumptions. When I was working in the UK in the 1970s, I was struck by the fact that advocates for the coal industry were able to show conclusively that it was cheaper than nuclear power, while representatives of the nuclear industry could show the opposite. Why? The basic reason is that there is no agreed basis for comparing the different costs incurred over a long period of time. This problem bedevils any comparison. It allows even intelligent people of good will to reach different conclusions, based on different legitimate assumptions. Those not of good will can also resort to a less legitimate basis for their sums to get the answer they (or their political masters) want.

A nuclear power station costs more to build than a coal-fired equivalent, which is not surprising. A power station boils water to drive a steam turbine. A nuclear reactor is a much more complicated way of heating water than a coal boiler. But a nuclear power station, once built and operating, costs less to fuel than the coal-burning system. So which costs less overall? It depends on the assumptions made about the future costs of fuel, as well as the way the calculation treats future costs and revenues. Economists have developed a technique called Discounted Cash Flow Analysis, to make it possible to compare future costs and income in equivalent present-day dollars. The discounting principle is a simple and obvious one. Generally, we would prefer to have $1000 now rather than the promise of $1000 in a year's time, even if the money was absolutely guaranteed. In fact, the entire business of lending money is based on the fact that we would so much prefer the $1000 now that we are prepared to pay $1100, or $1150 or even $1200, next year to have $1000 at the moment. Banks and other financial institutions lend us money on this principle. So the standard economic analysis brings all future costs and revenues into their equivalent current value, usually known as the Net Present Value (NPV).

If the interest rate was 10 per cent and I had $1000 now, it would be worth $1100 in a year's time. If I allowed the interest to accumulate, it would be worth $1210 after two years, $1331 after three years, and so on. Conversely, the net *present* value of $1000 to be given to me in a year's time is only $909, which is the amount I would need to have now for it to grow over a year into $1000; if I don't get it for two years, it is only worth $826 now, and if I don't get it for three years, its net present value is only $751. When large projects extending over several years are considered, the standard approach is to bring all the future costs and revenues into present-day values using this technique, often doing the calculation several times for different interest rates to see the impact on the project if the rates rise or fall. For example, I expect my household's investment in a solar water-heater to save us about $400 a year by reducing our electricity bill. Whereas a simple calculation would give $4000 in savings after ten years and $8000 if the solar panels last for twenty, the discounted values are about $2700 after ten years and about $3750 after twenty. In fact, the savings would only be $4000 in present value if the system lasted forever!

One of the problems with this type of analysis is that it leads to bizarre conclusions. Suppose you had a small farm which could either be managed to maintain its productivity, giving revenues of $40,000 a year far into the future, or used more intensively to produce $80,000 a year for seven years, after which it would be so degraded as to have no value. Which strategy should you adopt? Surely it's obvious that $40,000 a year for the next hundred years is worth more than $80,000 a year for seven years, wouldn't you think? Well, it isn't if you do a net present value calculation. $40,000 a year forever is worth $400,000 in present value, while $80,000 a year for seven years has a net present value of $428,400. So there is actually a small economic incentive to destroy the productivity of the land by farming intensively, rather than using it in a sustainable way.

This example shows that the assumptions underlying the economic analysis are crucial. If the future is heavily discounted, unsustainable practices appear to make economic sense. This problem is recognised by some economists, who include the "opportunity cost": the cost of the opportunities that the proposal would foreclose. In this case, the productive use of the farm would not be possible after the seven years of intensive exploitation, so the proposed course of action looks very foolish. Not so foolish that we can't find advocates for it, however! Precisely this sort of calculation has been used by resource economists to justify clear-felling forests, over-exploiting fisheries and slaughtering whale populations. In each case, the discounting calculation suggested that sustainable practice would yield a lower return in net present value terms. As a general rule, the practice of discounting future costs and benefits distorts decisions toward short-term profitability. Technologies which have a low capital cost and high running cost look better on this sort of analysis than alternatives with a higher initial cost, even if the subsequent running costs are much lower. Nuclear power suffers from this approach. So, ironically, do most forms of renewable energy. Coal-fired power stations require millions of tonnes of coal every year for their thirty years or so of operating, but those future costs are discounted.

Discounting does improve the apparent economics of nuclear power in one crucial way. At the end of the life of a power station, it has to be decommissioned. This is a much more complex task if you are pulling apart a nuclear reactor, because some parts of it will remain intensely radioactive. If we were to decide to build a nuclear power station now, if it took fifteen years to build and it was to operate for thirty-five years, the decommissioning would not begin for fifty years. The cost of that process, discounted to present values, would appear quite small – but it will represent a real cost for future generations. To quantify the impact of discounting, I did a simple calculation. A cost of $100 that will be incurred in fifty years' time has a net present value of $5; if it won't be paid for seventy years, the net present value is $1. So the discounting calculation would reduce a decommissioning bill of $1000 million to just ten or twenty million dollars, and declare it to be negligible. The same argument applies to waste. A nuclear power station produces radioactive waste which will need to be stored for extremely long periods, hundreds of thousands of years for a conventional current installation. Discounted to net present values, the cost of managing waste for hundreds or thousands of years is completely negligible. But again, it will be a real and substantial cost to future generations.

There are further complications. The cost per unit of delivered power depends critically on the "load factor" – the percentage of the possible total output the station actually delivers. This output is partly dependent on its reliability, since this determines how much of the time the power station is available to supply electricity. But it is also partly dependent on other factors that have nothing to do with the reliability of the power station: how much electricity is actually needed by the system, the way the load varies during the day and the way the system operators choose which power station to use from the range of options they have. If you want to make the economics of a proposed power station look attractive, you assume a high load factor, like 85 or 90 per cent. Some reliable power stations in well-managed grid systems do achieve that sort of figure, but many do not. If the calculation of the economics assumes a load factor of 90 per cent and the power station only achieves half of that, as has been the record of some UK reactors, the power will be twice as expensive.

So what did the Switkowski report say about the economics of nuclear power? The section on key findings says:

> Nuclear power is likely to be between 20 to 50 per cent more costly to produce than power from a new coal-fired plant at current fossil-fuel prices in Australia. This gap may close in the decades ahead, but nuclear power, and renewable energy sources, are only likely to become competitive in Australia in a system where the costs of greenhouse gas emissions are explicitly recognised. Even then, private investment in the first-built nuclear reactors may require some form of government support or directive.

The section on electricity generation claims:

> Nuclear power is the least-cost low-emission technology that can provide baseload power, is well established and can play a role in Australia's future generation mix.

Finally, the section on looking ahead says:

> There needs to be a level playing field for all energy-generating technologies to compete on a comparable whole-of-life basis. In a world of global greenhouse gas constraints, emissions pricing using market-based measures would provide the appropriate framework

for the market and investors to establish the optimal portfolio ... While carbon pricing could make nuclear power cost competitive on average, the first plants may need additional measures to kick-start the industry.

To summarise, the view of the taskforce is that introducing carbon pricing would make coal-fired power more expensive and go some way to closing the gap, but even assuming the likely price for carbon, nuclear power will probably still not be attractive to the commercial world. So a group that could not be considered even remotely hostile to the idea of nuclear power still concludes that it is not economically competitive and "additional measures" or "some form of government support" would be needed for reactors to be built. The most optimistic estimate it can produce is that nuclear would cost 20 per cent more than conventional power. This estimate is based on ambitious assumptions about the possible future level of government subsidy for nuclear power. The more realistic estimates in the report put the price of nuclear electricity at least 50 per cent higher than the present mix. Even these estimates do not include provision for insurance, since the commercial agencies are unwilling to cover the risks of the nuclear industry. So the public would effectively be the insurer, picking up the tab if anything goes wrong. This is the reason why many conservative observers are unsympathetic to the case for nuclear power: it simply does not make economic sense.

After the full report has been analysed, it is clear that there would need to be very generous public support to make nuclear power viable. The report cites US studies of the likely cost of new commercial nuclear power stations, which estimate wholesale electricity price in the range A$75–105 per megawatt-hour. By comparison, the report quotes Australian prices for coal as $30–40 per megawatt-hour, for gas turbines as $35–55, for wind or small hydro as about $55 and for solar thermal or biomass as about $70–120. Nuclear power is not marginally more expensive than coal or gas, it is at least double and perhaps three times the price, as well as much more expensive than the cheaper forms of renewable energy. Only solar cells directly converting sunlight into electricity, quoted at $120, and fossil-fuel power with carbon capture and storage, based on equally ambitious estimates of the future cost of that technology, are comparable with the US estimates of the likely cost of nuclear power. The report then assumes that the final price in Australia would be very much lower, in the region of $40–65, "if Australia becomes a late adopter of new generation reactors." This figure is remarkable for its breathtaking optimism. It assumes there will be future spectacular improvements in the design of a class of reactors that has not yet been built anywhere, roughly halving the price. Given that no such improvements have been achieved in the fifty-year history of the industry, one must conclude that there is no hard evidence for this belief. It really does justify the pejorative term some journalists use to describe any approach that doesn't support their political views: it truly is a "faith-based policy."

The report concedes that building nuclear reactors here is "likely to be 10–15 per cent more expensive than in the United States because Australia has neither nuclear power construction experience nor regulatory infrastructure," but applies this weighting to the dramatically lower price it has assumed rather than the hard-nosed estimates of US authorities. This is wishful thinking writ large. The Howard government and the commercial media accepted the report enthusiastically, proclaiming it as showing that nuclear power is on the threshold of being economical in Australia. The kindest conclusion is that none of those making this claim had actually read the report and absorbed the figures it cites. The claim is also based on an obvious logical contradiction. If Australia is to be "a late adopter of new generation reactors" and reap the promised economic rewards, we will have to wait until the reactors not yet designed are actually built and commissioned, some time in the 2020s at the earliest, before placing an order for a reactor which will then take another ten to fifteen years to be fulfilled. But this projected economic advantage is used to suggest that we should start building nuclear power stations immediately. That, of course, would totally preclude the advantage of being a "late adopter." In short, nuclear power is not economically viable in the foreseeable future, even if we do finally get an emissions trading scheme that puts a realistic price on carbon dioxide pollution.

The Switkowski report's economic analysis also includes some masterly sophistry. The report says, "Nuclear power is the least-cost low-emission technology that can provide baseload power, is well established and can play a role in Australia's future generation mix." Essentially the group decided that nuclear power is a low-emission technology, that it is "well established" and that cheaper renewable supply options are unable to provide baseload power, all of which are extremely contentious assertions, and then declared victory on the basis of this circular logic.

Baseload power is the level of electricity demand that exists twenty-four hours a day. Apart from traffic lights and subsidised aluminium smelters, there are relatively few power uses that operate around the clock. Demand for power is much lower at 3 a.m. than on winter evenings or, since recent aggressive marketing of airconditioning, on summer afternoons. Historically, the electricity industry has encouraged householders to use wasteful technology like storage electric water heaters by offering "off-peak tariffs," the purpose of which is to shift energy demand to hours when the system has plenty of spare capacity. The policy was developed to improve the operating performance of inflexible large power stations that are only cost-effective if they run night and day. It artificially inflates the baseload. The real demand for baseload power in Australia is relatively small. The biggest user of it is the aluminium industry, which is so heavily subsidised from the public purse, employs so few people and is so largely overseas-owned that an Australia Institute study concluded it would cost us less if we closed it down and paid every worker in the industry $100,000 a year to go fishing! It is also true that several countries use renewables for their baseload: New Zealand, Norway and Iceland are three countries with electricity systems where almost all of the power comes from renewables, mainly hydro and geothermal. While there is a widespread misapprehension that renewables like wind can't be expanded beyond about 10 per cent of an electricity grid without risk of instability, there are certainly examples in Australia of small-scale power grids that go well beyond that without any problems. Birdsville gets about 70 per cent of its power from the heat in artesian bore water, while the WA area of Esperance at peak times gets almost 70 per cent of its electricity from the local wind farm (although wind power only fulfils about 22 per cent of the total demand). Even if one accepts the questionable claims that the present baseload is irreducible and that nuclear power is a low-emission technology, it is certainly not the only way of supplying future electricity capacity. And the claim that it is the "least-cost" option is just wrong on any reasonable interpretation of the data cited in the Switkowski report. I discuss in more detail in the final section of this essay the clean alternatives for baseload power.

Other parts of the nuclear lobby resort to extraordinarily contorted figures to justify their faith in the economics of nuclear power. A research group at the University of Melbourne, in an aggressively pro-nuclear report, concedes that the construction of nuclear power stations has been marked by delays and huge cost over-runs. It then says, "Westinghouse claims its Advanced PWR reactor, the AP1000, will cost US$1400 per kilowatt for the first reactor and fall to US$1000 per kilowatt for subsequent reactors. They also claim that these will be ready for electricity production three years after first pouring concrete." Noting that the actual experience in the US involved construction costs of up to US$5000 per kilowatt and much longer completion times, the report goes on to conclude its section on economics by saying, "If the AP1000 lives up to its promises of $1000 per kilowatt construction cost and three-year construction time, it will provide cheaper electricity than any other fossil fuel based generating facility, including Australian coal power, even with no sequestration charges." Well, yes; if the cost of future power stations was only one-fifth of that of past experience, and the construction time was between a half and a quarter that in the past, the economics would look a lot better. That is a bit like saying that we should take no account of the fact that my score in my most recent round of golf was about ninety, and indeed all my recent scores have been around that figure; I would only have to get it down to the mid-sixties and I could win next year's Australian Open, so I expect government support for my entry.

A crucial political consideration is the public hostility to reactors. In Europe and the US, this has caused protracted delays in construction. Since there is similar hostility in Australia, it is naive or dishonest to base economic assessment on the assumption that reactors could be built in three years. At a recent Adelaide conference, the editor of the industry journal *Nucleonics Weekly* gave a paper warning of excessive optimism about construction times and costs. Mark Hibbs said Westinghouse had lost "several hundred million dollars" on a new reactor project in Finland, while the design for two proposed reactors in Taiwan is only about 65 per cent complete, eleven years after the signing of contracts with two US firms. Some of the delays can be attributed to political divisions and consequent legal battles, but the costs of some components have increased by as much as a factor of six. He warned that the designs for new reactors are still on the drawing board, so there is real uncertainty about the actual costs. Hibbs also said that it is unclear whether manufacturers would even be interested in building one or two reactors in a country with no infrastructure or past experience of nuclear power. In other words, it may not even be possible to interest the commercial nuclear industry in building an Australian reactor; even if a reactor were feasible, no one can really say how expensive it would be. By contrast, we have experience of building wind turbines and installing solar hot water systems, so we have both the necessary skills and confidence in the economics.

The Switkowski report says that at least ten and possibly fifteen years would be a realistic time-scale for building *one* nuclear power station in Australia. It would take more time still to "pay back" the energy used in construction and fuelling, so it would take fifteen to twenty years for any such station to make any contribution to cutting greenhouse pollution.

Fifteen to twenty *months* is a more realistic time scale for large-scale renewables. Global warming is an urgent problem that demands a concerted response now, not a half-baked response after 2020.

Besides, the scale of the potential impact on our greenhouse pollution is not impressive. The most aggressive pro-nuclear scenario analysed by the report projected twenty-five nuclear reactors dotted around the nation, but this would only have the potential to reduce the *growth* in our greenhouse pollution by between 8 and 18 per cent. Even the higher figure would be a miserable attempt to meet our greenhouse responsibilities. The science shows we need to cut greenhouse pollution by at least 60 per cent, and probably by more like 80–90 per cent. Also, the nuclear option would involve a massive release of carbon dioxide in the next few decades, building and fuelling reactors, just at the time we should be cutting back. It doesn't make sense as a response to climate change.

Reactor safety is also an issue. Australians are understandably worried about the prospect of accidents. All the report can say is that the promised new generation reactors, still on the drawing board, should be safer. This is an expression of hope rather than an evidence-based assessment. These assurances come from the same group of people who told us that the existing reactors were safe and dismissed the Chernobyl disaster on the grounds that it was a primitive Russian design. The Switkowski report says little about where nuclear power stations would actually be sited. This is clearly a critical political issue, as demonstrated by the haste with which anyone trying to clarify the location of potential sites is branded a scare-monger. Power stations should be placed near centres of high electricity demand, which means either the large cities or our subsidised aluminium smelters. They also need cooling water, which means they will need to be on coastal sites, given the water crisis caused by climate change. This is a serious issue; some nuclear power stations were unable to generate electricity in the 2006 northern hemisphere summer because of a shortage of cooling water! The Australia Institute identified as suitable sites various coastal locations near the major mainland cities: Sydney,

Melbourne, Brisbane, Perth and Adelaide. Such findings lead some observers to suggest nuclear power is politically impossible in Australia, since they believe no government that proposed building a reactor on a coastal site near a major city would be re-elected. That is probably true in the present political climate, but governments have a great capacity to try to change public views; the present Commonwealth government is spending hundreds of millions of dollars of our money trying to persuade us of the benefits of some of their unpopular policies. I discuss later the likelihood of public views about nuclear power being changed by government "education" campaigns.

The Switkowski report accurately concludes that disposal of high-level waste is "an issue" in most countries using nuclear power. Until the problem is resolved, it is irresponsible to produce more waste. It is contributing to a problem that currently does not have a solution, dumping it on future generations to resolve. On these grounds, it fails one of the tests of sustainability: inter-generational equity. Such an approach would not be morally defensible, even if we did not have alternatives; but we do.

We should also worry about the possible impact of our choices on proliferation of nuclear weapons. The report claims that "increased Australian involvement in the nuclear fuel cycle would not change the risks," but this seems naive. Iran's neighbours are nervous about an energy-rich country that clearly does not need nuclear power embracing this technology, suspecting its real motives. It would be equally understandable if our neighbours drew the same conclusion. Any agitators wishing to spread the idea that Australia was planning to build nuclear weapons would find enough historical urgings from nuclear technocrats to make their case appear credible. The former head of the then Australian Atomic Energy Commission, the late Sir Philip Baxter, was an unashamed advocate of developing a nuclear weapons capacity. If our government were so foolish as to go down this path, it would dramatically increase the risk of proliferation in our region. Visiting Australia recently, former US vice-president Al Gore observed that every problem of weapons proliferation during his eight years in the White House arose from a civilian nuclear program. In our area China, India and Pakistan all developed nuclear weapons in association with nuclear energy. The same could be said of Israel, while in the UK the nuclear energy industry began as a smoke-screen to conceal the real agenda of building bombs. The more countries have nuclear weapons, the more certain it becomes that one will become deluded enough or desperate enough to use them.

After it has been mined, uranium undergoes a process called enrichment. The reason for this is that the naturally occurring mineral is a mixture of two slightly different forms known as isotopes: Uranium 235 and Uranium 238. U235 is much more radioactive than U238, but the natural mineral consists mostly of the less active form. So an industrial processing technique was developed, initially as part of the Manhattan Project to build the nuclear weapons which were dropped on Japan in 1945, to "enrich" the uranium by increasing the proportion of the more reactive isotopes. The process requires huge amounts of energy and is very expensive. Increasing the fraction of U235 to about 10 to 20 per cent is normal practice when the uranium is to be used in a nuclear power station, while enriching it to 80 or 90 per cent makes it suitable for nuclear weapons. Because the same approach can be used either to produce reactor fuel or weapons-grade material, uranium enrichment is seen as a very sensitive technology. There is at present considerable unease about Iran developing the capacity for enrichment, largely because of the fear that the technology could easily be applied to develop an Iranian bomb.

In mid-June 2007, the ABC's Investigative Unit unearthed an amazing story about negotiations between the Commonwealth government and representatives of a company called Nuclear Fuel Australia to allow uranium enrichment in Australia, even to the point of identifying two possible sites, one in South Australia and one not far from where I live in southeast Queensland. The story was a revelation because even the Switkowski group had poured cold water on the idea of Australian involvement in enrichment, though this has been a dream of the nuclear lobby for at least thirty years. At the National Press Club, discussing his report's conclusions, Dr Switkowski said:

> The technology for enrichment is tightly held by a small number of companies which are currently adding modern centrifuge plants.

Thus there appears little need for new capacity in these processes for some time. On the other hand, the forecast dramatic rate of growth of nuclear power in the Asian region may eventually support investment in a regional enrichment plant.

Enrichment is a proliferation-sensitive technology and the subject of international scrutiny.

Australia has a strong non-proliferation record and all fuel cycle activities are covered by Australia's safeguards agreement with the IAEA. Other than normal commercial considerations, any decision to get into value-adding at the enrichment stage would further require a careful assessment of security implications for Australia and the management of international perceptions. That being said, in the panel's view, these issues are not insurmountable.

This really is damning the idea with very faint praise indeed: "little need for new capacity ... for some time," "the forecast dramatic rate of growth of nuclear power in the Asian region may eventually support investment," as well as admitting that enrichment is "a proliferation-sensitive technology" that would require "careful assessment of security implications for Australia and the management of international perceptions." The Switkowski report also noted that there are "high barriers to entry" into the enrichment activity, and that supply of enriched uranium is expected to exceed demand until at least the middle of the next decade. BHP Billiton, which mines uranium at Roxby Downs, has explicitly ruled out the possibility of enrichment. There is no serious analysis in the report of the economics of enrichment, but Dr Alan Roberts has noted that it is not strictly commercial anywhere in the world, being subsidised at least partly because the technique is also used to produce weapons-grade material. As Roberts concisely summarised the matter, "Wherever uranium is enriched, the taxpayers are impoverished."

Although the idea that Australia could enrich uranium does not appear to make any sense, either commercially or politically, John Howard gave a wink to the nuclear lobby when he set in train the public discussion. On ABC Radio in June 2006, he said:

> Well, we should look at uranium enrichment. Of course we should. We as a nation for generations have lamented the fact that we had the finest wool in the world, but we sent it overseas to be processed. I don't want ... if there's a viable economic, safe alternative, I don't want the same thing to be said in future generations about our uranium.

This is an extraordinary statement because the whole economic thrust of the Howard government has been to export ever-increasing volumes of low-value commodities such as iron, coal, wool and cotton. Value-adding in these areas has been systematically run down with the closure of steelworks and the collapse of the textiles and clothing industries, all driven by government policies – yet suddenly, in the case of uranium, we should not just be exporting the raw material but also considering further involvement in processing. Pro-nuclear MPs ran a similar line in the House of Representatives committee which reported on the nuclear issue in 2006. The report said:

> Federal and state government decisions over the past thirty-five years have led to the abandonment of several opportunities to develop industries based on upgrading Australia's uranium resources for export. Perhaps the most significant of these missed opportunities involved a proposal to develop a commercial uranium enrichment industry in Australia by a consortium of Australian companies, the Uranium Enrichment Group of Australia – BHP, CSR, Peko-Wallsend and WMC – in the early 1980s. This proposal was terminated following a change of federal government.
>
> By the mid-1980s, the Australian Atomic Energy Commission (AAEC) had accrued twenty years of experience with uranium enrichment technology. The Commission had by then invested

some $100 million on enrichment research alone. This knowledge and expertise was lost following the federal government's direction that the Commission and its successor agency, the Australian Nuclear Science and Technology Organisation (ANSTO), abandon enrichment and other fuel cycle research.

Perhaps this is where the dog is buried, as the Russian saying goes. The proposals to develop uranium enrichment were put forward in the days of the Fraser government, when John Howard was treasurer. They were even an election issue; I remember attending a public meeting in Caboolture Town Hall in early 1983 when about 600 angry residents voiced their disapproval of the scheme to inflict uranium enrichment on their town (in an uncanny echo, the ABC's Investigative Unit discovered in 2007 that a site near Caboolture is again being considered for an enrichment plant). Is John Howard's support for enrichment an attempt to justify the $100 million of public funds he wasted on this nuclear pipe-dream on his previous watch? There has been no discussion of the huge energy demand of such a plant, nor of the cost of transporting millions of tonnes of uranium oxide to the plant and similar quantities of depleted uranium and other waste to some disposal site, nor of the considerable environmental impact of such an operation, especially if conducted in such a densely populated area as south-east Queensland. The House of Representatives committee even made the amazing claim that this waste would be valuable! The report said, "The operation of an enrichment plant will produce depleted uranium of an amount some seven times greater than the enriched uranium produced. This depleted uranium would constitute a tremendous energy asset for future use here and/or overseas (for use in breeder reactors)."

This is ingenious in the extreme. Don't worry about the problem of a volume of radioactive waste seven times greater than the finished product, don't try to manage it or convince the local community that it would be acceptable, convince yourself instead that the idea of the breeder reactor, abandoned by all but the most fanatical nuclear zealots in the northern hemisphere, will be resurrected and provide a use for the waste. This byproduct is, by definition, depleted in the process of producing enriched uranium, so it actually contains less of the highly reactive isotope than the ore from which it was processed, but it will still "constitute a tremendous energy asset." That is a bit like describing the spoil heaps near old coalmines as an energy asset rather than an environmental risk.

Confirming the pro-enrichment line in Canberra now, John Carlson, the head of the federal government's Australian Safeguards and Non-Proliferation Office, said he believes uranium enrichment is government policy. Defending a proposal to export uranium to Russia, Carlson said, "It's too early to say whether we will or we will not participate in the Russian [enrichment] centre but I think our principle is to have enrichment in Australia in the future." This immediately led the ALP environment spokesman, Peter Garrett, to question where the government's uranium policy was heading: "Since a senior government official has now admitted that we will have nuclear enrichment in Australia, we also need to know what discussions and commitments were given to Russian officials or to any other countries concerning Australia being used as a dumping ground for the world's nuclear waste." While there is no necessary link to other stages of the nuclear industry such as waste management, Garrett was correct in pointing out that the government seems to be saying things internally that it would not want the public to hear. The Howard government would presumably have preferred that all this shady dealing had not become public before the election. Interviewed on the 7.30 *Report*, a clearly uncomfortable Dr Clarence Hardy of the Australian Nuclear Association conceded that the result of the election was very likely to determine whether the proposal would go ahead. That sounded remarkably close to an admission that the Howard government has secretly agreed to support the idea if re-elected.

THE GATHERING THREAT

After the US used nuclear weapons against Japan, most people were so shocked that they believed a fundamental task of the new United Nations organisation was to curb the spread of such weapons. When the US government declined to share the nuclear technology that had been developed during World War II with the UK, that country decided secretly to develop its "independent nuclear deterrent." The Calder Hall reactor was designed to produce fissile material for the British bomb, but as a smokescreen it was also set up to produce a small amount of electricity for the power grid. France, China and the USSR (of which Russia was by far the largest fragment) also developed nuclear weapons. At the height of the Cold War, something like 50,000 nuclear warheads were poised ready for use and the strategic analysts used obscene terms like "overkill" to reflect the fact that the nuclear powers had the capacity to kill everyone on Earth. Most leaders recognised that the situation was extremely perilous. The rational ones among them could see that the weapons could not be used without provoking an appalling retaliation, although some in the military persisted in the delusion that it might be possible to "win" a nuclear war, even if whole cities were lost and millions died. The most extreme example of this lunacy that I have seen was an article on the front page of the *San Francisco Chronicle* in 1984, proudly stating that the US Navy was "ready for World War Four." The story revealed that a Poseidon submarine with 600 nuclear weapons was permanently stationed under the polar ice-cap to guard against the possibility that the US might have no weapons left after a full-scale nuclear war. "We want the Russkies to know that we would still be able to make the rubble bounce a bit," a military leader was quoted as saying. Bizarre thinking of this kind gave rise to an international movement for nuclear disarmament, but the wishes of those under threat were blithely ignored by the military and political leaders of nations with nuclear weapons.

There had been an earlier move to remove the Sword of Damocles

suspended above civilisation: the Nuclear Non-Proliferation Treaty, usually known as the NPT. According to this 1970 agreement, the technology
for peaceful use of nuclear energy would be made freely available, allowing all nations who wanted to use it for power generation to do so. At the
same time, a concerted effort would be made to rid the world of nuclear
weapons, with the nations holding weapons agreeing to reduce their
stockpiles and those without weapons refraining from their development. The problem is that the plan hasn't worked. There are still fearsome
stockpiles, amounting to thousands of weapons in the case of the US and
Russia, while nobody is quite sure how secure are the weapons now in
Russia and other parts of the former USSR. Both the US and China are still
developing their abilities to use nuclear weapons, with the US military
talking about "mini-nukes" – as if they were cute and cuddly – while
China brazenly expands its capacity to use rockets to deliver weapons. The
Bush administration's aggressive promotion of its missile defence system
is further adding to the tension. In the absence of serious moves to disarmament by the original nuclear weapons states, India, Pakistan and
Israel have openly developed their own weapons, while both North Korea
and Iran are heavily involved in nuclear technology. It was recently
reported that Pakistan is commissioning a new reactor that looks suspi-
ciously like one designed to allow production of plutonium for nuclear
weapons. It has also been reported that Israeli intelligence believes Iran
could have a bomb within two years.

Though the Australian Safeguards and Non-Proliferation Office (ASNO)
is the agency officially concerned with these issues on our behalf, it
appears to be quite sanguine about the future. Certainly it appears to be
happy with the status quo. The former diplomat Professor Richard
Broinowski reminded the House of Representatives Committee that there
are double standards for the parties to the NPT, saying:

> Under article VI, the weapons states are supposed to reduce, and
> then do away with, their arsenals [in return] for the non-nuclear

weapons states saying, "We will not possess, develop or acquire nuclear weapons." In my view, we are going to see one or two extra nuclear states every year because they are absolutely sick and tired of having to follow their part of the bargain while the superpower and the other nuclear weapons states have no intention of reducing their armaments. Indeed, the United States has new programs to make new weapons.

ASNO responded by saying that a closer examination of the actual obligations is required because the disarmament provisions are more complex than many critics appreciate. Article VI of the NPT doesn't actually require the weapons states to disarm but to "pursue negotiations in good faith on effective measures relating to cessation of the nuclear arms race at an early date and to nuclear disarmament, and on a treaty on general and complete disarmament under strict and effective international control." Since this disarmament commitment involves all parties, not just the weapons states, ASNO argued that it does not place all the onus on the weapons states. It also said that both the US and Russia "have reduced the numbers of deployed warheads from 10,000 each in 1991 to 6000 each in 2002, and are proceeding to levels of between 1700 and 2200 by 2012 in accordance with commitments contained in the 2002 Moscow Treaty on Strategic Offensive Reductions." ASNO – who seem as interested in promoting the use of uranium as in ensuring that the safeguards on its export are effective – have thus maintained that the weapons states don't actually have to disarm, they only have to negotiate in good faith – but it isn't clear they are even doing that. In any case, it is not an argument that should give us any great comfort. The fundamental point is that the non-weapons nations were so angry about the lack of progress that the 2005 NPT review conference ended in disarray and the most likely eventuality is a further spread of nuclear weapons.

Differences of interpretation bedevil the NPT. Article IV proclaims "the inalienable right of all Parties to the Treaty to develop research, production and use of nuclear energy for peaceful purposes without discrimination." Iran's leaders believe that their contentious nuclear initiatives are entirely justified under that provision, and some local observers, such as Professor Broinowski, agree:

> Iran right now are acting perfectly legally under the NPT and the extended protocol in developing an enrichment plant. Indeed, they are encouraged to do that by the terms of the NPT and its extensions. Yet that could lead immediately to weapons-grade plutonium or uranium being developed in Iran. All you have got to do is go beyond a 20 per cent U235 to up to 90 per cent and it is the same process.

The Iran example demonstrates a fundamental tension in the NPT regime, which seeks to encourage peaceful use of nuclear technology while restraining its military Siamese twin. The pro-nuclear lobby views the present situation not as alarming, but as a good outcome, by pointing to even worse alternatives and inviting us to feel relieved. Ian Hore-Lacy heads the Uranium Information Centre, which is funded by the mining industry to distribute pro-nuclear material. He told the parliamentary committee, when it expressed concern that nuclear power might lead to weapons proliferation:

> I think there would probably be two or three times as many weapons states now if there were no civil nuclear power, because the Nuclear Non-Proliferation Treaty has had this tradeoff of technical assistance for the development of civil power on the basis that people stood back from the possibility, and eschewed the possibility, of developing weapons. In the 1960s there were a number of reputable estimates that by the turn of the century there would be at least thirty, probably thirty-five, nuclear weapons states. Now we have five official ones, we have three unofficial ones ... I think that is an extraordinarily good result.

I suppose it is always true that things could be worse, but you have to be wearing very superior rose-coloured glasses to see the present situation as a demonstration of the success of the Nuclear Non-Proliferation Treaty. The same line was repeated by the House of Representatives committee, which described the fact that the original five nuclear-armed states have only been joined by a few more as "clearly a tremendous achievement"! As Al Gore said about climate change, it is remarkable how hard it is for people to see the glaringly obvious when their job depends on them not seeing it. The fundamental problem has been stated by Mohamed ElBaradei: "As long as some countries place strategic reliance on nuclear weapons as a deterrent, other countries will emulate them. We cannot delude ourselves into thinking otherwise." The extraordinary contrary position espoused by the Australian Safeguards and Non-Proliferation Office is that disarmament has stalled only because nuclear proliferation is such a pressing issue these days! In other words, the weapons states are justified in not disarming because a small number of the outsiders have now responded to the lack of progress by starting moves to join the nuclear club, making the world a more dangerous place in which nuclear weapons are necessary. If this wasn't such a serious issue, that would be a laughable stance; it certainly brings little credit to our national government.

The House of Representatives committee cited Dr ElBaradei's view that three developments in recent decades have transformed the nuclear security landscape: "the emergence of clandestine nuclear supply networks; the spread of fuel cycle technologies; and renewed efforts by a few countries and some terrorist groups to acquire nuclear weapons." For ElBaradei, these developments have highlighted "several vulnerabilities in the non-proliferation regime, including the limitations on the IAEA's verification authority, control of proliferation-sensitive technologies, and the IAEA's technical capability to detect undeclared nuclear activities." This is a remarkably honest statement from the head of a body which usually sees its role as promoting the use of nuclear technology and which cannot be accused of being hostile to the nuclear industry.

In 2006, *An Illusion of Protection*, a report prepared for the Australian Conservation Foundation and the Medical Association for the Prevention of War, was published. It drew together another set of statements by ElBaradei:

> The IAEA's Illicit Trafficking Database has, in the past decade, recorded more than 650 cases that involve efforts to smuggle [nuclear and radioactive] materials.
>
> Today, out of the 189 countries that are party to the NPT, 118 still do not have additional protocols in force.
>
> IAEA verification today operates on an annual budget of about $100 million – a budget comparable to that of a local police department. With these resources, we oversee approximately 900 nuclear facilities in seventy-one countries ... we are clearly operating on a shoestring budget.
>
> If a country with a full nuclear fuel cycle decides to break away from its non-proliferation commitments, a nuclear weapon could be only months away.
>
> We are in a race against time.

The "additional protocols" referred to by Dr ElBaradei are the extra safeguards added to the NPT when it was discovered in 1991 that Iraq had secretly made significant progress toward acquiring nuclear weapons. It was the recognition that the safeguards regime could be so easily circumvented that led to their development. Yet, as Dr ElBaradei noted, fifteen years later they have only been adopted by a minority of the NPT signatories, some seventy-seven nations.

The 2006 report of the Weapons of Mass Destruction Commission, chaired by former IAEA head Dr Hans Blix, reinforced the diagnosis:

> The Commission rejects the suggestion that nuclear weapons in the hands of some pose no threat, while in the hands of others they place the world in mortal jeopardy. The three major challenges the

world now confronts – existing weapons, further proliferation and terrorism – are interlinked politically and also practically: the larger the existing stocks, the greater the danger of leakage and misuse.

What of Australia's role in all of this? The *Illusion of Protection* report concludes that "there is a serious and unavoidable risk that Australian uranium exports to China will directly or indirectly support Chinese nuclear weapons manufacture, and potentially nuclear weapons proliferation in other countries."

In a foreword to the report, Dr Frank Barnaby, a nuclear physicist and weapons scientist in the UK, endorses the view that the safeguards system has such fundamental flaws "that the most prudent and responsible position is to oppose the mining and export of uranium." The world, he says, "would be a much safer place if the Australian government acted on this advice."

The international safeguards system is administered by the International Atomic Energy Agency. Critics argue that the IAEA is compromised as a regulator by its broad aim of encouraging use of nuclear energy, but that is another story. As the Switkowski report summarised the situation:

> Safeguards are a system of technical measures – including inspections, measurements and information analysis – through which the IAEA can verify that a country is following its international commitments to not use nuclear programs for nuclear weapons purposes.

The Australian government requires that nations buying our uranium must be parties to the NPT, must have bilateral safeguards agreements with Australia and (if they don't already have nuclear weapons) must have accepted the IAEA's additional protocols. The Australian bilateral safeguards agreements specify that our uranium should only be used for peaceful purposes, should be covered by IAEA safeguards and should not be transferred to a third party for enrichment beyond 20 per cent of U235 or for reprocessing without prior Australian consent.

As the journalist Robert Milliken argued in 1980, Australia's export safeguards have so many weaknesses that together they give only an illusion of protection. Since then, it could even be argued that the safeguards have been diluted whenever they looked like preventing a commercial deal. The Fraser government relaxed the safeguards to allow export of uranium to France, even though it was testing nuclear weapons in the South Pacific and had not signed the NPT. There used to be a requirement that prior written consent would be needed before nuclear fuel could be re-processed, an operation that can be used to extract plutonium for weapons production; that was diluted to a provision known as "programmatic consent" which actually allows re-processing of fuel elements. The Australian safeguards allow what is called "flag-swapping"; essentially this says that we aren't worried whether our uranium is used for weapons production or not, as long as a volume of uranium equivalent to our sales is kept solely for peaceful use. As the *Illusion of Protection* report says, "Australia's system of safeguards is a book-keeping exercise that relies upon the importing state to adhere to the material accountancy system. This can be murky in the case of nuclear weapons states because of the clear and proven linkages between civil and military facilities …"

Only last year it was revealed that an Australian company had sold uranium to Taiwan, by way of an intermediary in the US, thus contravening two provisions of the safeguards: onward sale to a third party and supply to a state which is not a signatory of the NPT. The weakness of our export safeguards raises the real possibility that Australian sales of uranium to China will allow that country to use its own uranium to produce nuclear weapons. So we have now supplied fissile material to both sides of the very tense dispute about the future of Taiwan.

On 14 August 2007, the decision was made to sell uranium to India, which is also not a signatory to the NPT. Again, the government seems to have looked for ways to justify further dilution of the safeguards to allow the sale, confirming the widespread view that the only purpose of the safeguards is to deflect *political* fall-out. Predictably, a Pakistani minister claimed that supplying Australian uranium to India would allow that country to use its own uranium to produce more weapons, on which basis he called for Pakistan also to be sold our uranium to maintain the balance. If the aim is simply to make more money, it is profitable to supply both sides of a nuclear arms race, but this is clearly an irresponsible policy. As the Fox Report stated more than thirty years ago, we are inadvertently contributing to the possible proliferation of nuclear weapons by exporting uranium.

Terrorism

While the public discussion has been dominated by the fear that rogue states might join the nuclear weapons club, the International Atomic Energy Agency has also expressed concern about the risk of terrorist organisations obtaining fissile material. The US president, George W. Bush, recently claimed that intelligence agencies had foiled a plot to detonate a "dirty bomb," a device which would use conventional explosives to disperse radioactive material and contaminate large areas of a city. It should be obvious that this is a special class of risk that only applies to nuclear power. There is no danger terrorist organisations will do appalling things with wind turbine blades or solar panels. Several reports have found that there is already a large amount of fissile material unaccounted for by the standard book-keeping exercises; the authorities just do not know where it is. The optimistic view is that it is simply lying around in quantities too small to be dangerous; the pessimistic view sees it as entirely possible for Al Qaeda or other extremist organisations to collect radioactive material with the aim of producing either a primitive nuclear weapon or a dirty bomb. Either would have truly dreadful consequences. The possibility that greed or malice could lead to such an event is a compelling argument against further increasing the amount of fissile material moving around the world.

Interviewed on *Lateline* early in July this year, BBC correspondent and author Gordon Corera said that Osama bin Laden had asked Pakistani nuclear scientists to help him get a nuclear weapon some time ago. According to Corera, it is becoming more likely that Al Qaeda will get a bomb as nuclear technology spreads to more countries. This underlines the problem that arises from fundamentalist religious views. The men who flew aircraft into the World Trade Center and who detonated suicide bombs on trains in London and Madrid believed that they would go directly to paradise, securing various benefits for themselves as holy martyrs, as well as guaranteeing salvation for large numbers of their friends and relatives. They had no fear of death and believed they were doing holy work by ridding the Earth of unbelievers. How much more fulfilled would they feel if they were capable of eliminating hundreds of thousands with a nuclear weapon? While that is a pathology specific to Islamic fundamentalism, others have observed that there are devout Christians within shouting distance of the Oval Office who believe global climate change or even nuclear war could be seen as heralding the Second Coming and their ascent to paradise. The general point, as the author of The End of Faith, Sam Harris, argues, is that these sorts of beliefs have unimaginable consequences when combined with real "weapons of mass destruction." Until now, the biggest restraint on the use of nuclear weapons by national governments has been the fear of retaliation. But this does not apply to terrorists, who have no territory to defend. They are more likely to use dirty bombs or even crude nuclear weapons than national leaders, even the more desperate or disturbed ones.

Dr Trevor Findlay is one who has warned of the risks. A former diplomat who was a member of the Australian delegation to the Conference on Disarmament in Geneva, Findlay spent seven years directing the London-based Verification Research, Training and Information Centre. Addressing a recent seminar in Adelaide on the subject of uranium, he said that increased global use of nuclear energy would demand a concerted response, focused not just on national obligations and action but on "enhanced global governance." His assessment was that the existing treaty provisions do not appear adequate even to deal with the present situation, let alone a future in which there would be a major expansion in the use of nuclear materials. Such a future would pose huge challenges for regulatory, monitoring and verification regimes. His words inevitably provoked recollection of the despairing call by Dr ElBaradei for more resources, when he pointed out that the IAEA is supposed to monitor over 900 nuclear installations with the budget of a local police force.

Similar conclusions were reached by the Oxford Research Group, an expert body concerned with security issues. In a report called *Too Hot to Handle? The Future of Civil Nuclear Power*, the ORG analysed the feasibility and the risk of using nuclear power to contribute to the problem of global climate change. It concluded that to play a significant role in curbing greenhouse pollution, nuclear power would have to provide at least one-third of the world's electricity by 2075. This would mean building nuclear power stations at a rate of four a month. This is not a credible rate of construction, the ORG found, but if it could somehow be achieved, it would also have a dramatic effect on proliferation. Such expansion in nuclear power would mean using much poorer grades of ore, thus undoing the greenhouse benefits, or else developing breeder reactors to use the nuclear fuel more efficiently. But the breeder path, even if it were technically feasible, would have disastrous implications for security. The ORG calculated that the proposed expansion would lead to 4000 tonnes of plutonium being processed into reactor fuel every year. That annual rate of production is about twenty times the current military stockpile of plutonium. In its report, it assessed that such a program would make it very probable that fissile material "would end up in the wrong hands" and be used to make either sophisticated weapons or dirty bombs. So, the ORG concluded, "A world-wide nuclear renaissance is beyond the capacity of the nuclear industry to deliver and would stretch to breaking point the capacity of the IAEA to monitor and safeguard civil nuclear power." This seems an obvious conclusion, given that the IAEA admits that it is already struggling to keep the weapons genie in the bottle.

It is sometimes suggested that the solution to the proliferation and waste problems might be for the producers of uranium to take responsibility for it. In other words, Australia would not just mine and export uranium, trusting the purchasers to use it responsibly, but would process the mineral into fuel rods and lease them to users on condition they be returned for disposal. That idea re-emerges from time to time, but it has little support even within the uranium industry. I suspect the producers of uranium believe they have conditional public support while they are simply exporting the mineral and making money, a fraction of which flows into the local economy, but that support would evaporate if the consequence was the return of radioactive waste to be managed for geological time. Thus, the South Australian state government has been vigorously promoting an expansion of uranium mining and export. But it also campaigned to stop the Commonwealth imposing a low-level radioactive waste dump on the state. The latter stance clearly had community support, tacitly accepted by the Howard government when they withdrew the proposal after alarmed responses from local Liberal MPs before the 2004 election. So the SA government is totally opposed to dumping waste in the state at the same time as it is a fervent promoter of uranium mining and export, which has the inevitable consequence of producing more radioactive waste. Environmental groups opposed to uranium mining are understandably against involvement in other stages of the nuclear industry. Even the Australian Safeguards and Proliferation Office has opposed the idea, saying that it would not be practical to manufacture fuel assemblies for Australia's uranium purchasers, who between them are claimed by ASNO to use "some sixty different reactor models" – an alarming thought in itself.

Pro-nuclear advocates often point to the relatively small volume of such waste by comparison with the spoil from coalmining. While they have a point in saying that the volume of radioactive waste is relatively small, it is not an honest comparison because coal waste does not have to be isolated from the biosphere for hundreds of thousands of years.

The other traditional argument is that the problem is, at least in principle, capable of being solved, so we should not worry about it. This is also specious. Aircraft are, in principle, able to fly safely. There is no technical reason why the flight I intend to board next week should not safely reach its destination. The reason we have a Civil Aviation Safety Authority is that technical capacity is not enough; the human systems also have to operate properly. From time to time, aircraft do crash. Sometimes the cause is a mechanical failure, but far more often the accident is the result of human error, either in flying the aircraft or in such support services as air traffic control or maintenance. Most major airlines use the same types of aircraft, so the large differences in accident rates are mainly a function of social organisation. To ensure that the system operates safely, we need to have confidence in the social organisation of the task as well as accepting that the technology is adequate.

While I was working in the UK, the nuclear authority there discussed the possibility of encasing radioactive waste in blocks of glass for disposal. As a materials scientist, I was horrified by this proposal. Glass is not really a solid; it is what is technically called a super-cooled liquid, meaning it is not stable over centuries at room temperature. It seemed entirely fanciful to suppose that glass could be subjected to the heat and ionising radiation of nuclear waste and still retain its structural integrity for thousands of years. Nuclear authorities, then and now, usually try to dismiss this concern by saying that the glass would only be one layer of defence.

The report prepared by the Australian Science and Technology Council for the Hawke government in the 1980s argued that waste would be put in glass blocks inside metal containers surrounded by clay and deposited in stable geological formations, such as old salt mines. This would certainly be a better approach, because the system would only fail and release radioactivity if all the layers were breached by some catastrophic event. The probability of contamination would be reduced to a lower level. The question of whether it would be acceptably low is a social value judgment. The social issue should not be discounted. Given the hostility of South Australian voters to the proposal to store low-level waste, including such items as gloves and lab coats lightly contaminated by use in hospitals or university research facilities, it is difficult to see why any community would want to accept the much more highly radioactive waste from a nuclear power station.

Nor is this a problem peculiar to Australia. The fundamental reason why no "high-level" nuclear waste has yet been processed for final disposal anywhere in the world is that every such suggestion has aroused fierce opposition from local communities. For decades, the US nuclear industry had been planning to store their waste in the state of Nevada, at a site with the wonderfully appropriate name of Yucca Mountain. But local opposition has finally ruled out that site; even Nevada, which runs on gambling, isn't prepared to bet on nuclear waste. That leaves the USA without a plan B.

I fear we are being lined up to receive the waste that nobody wants in the US. When asked in 2006 if we were being set up to be the world's nuclear waste dump, Mr Howard denied the claim, but did say that if we exported uranium to a country, we might be seen as having some responsibility for the waste that was produced by its use. Since we sell uranium to the US, that comment could be used to justify taking American waste. Dr Helen Caldicott suggested in a television debate that we should be alarmed about this possibility. She noted that the Federal Council of the Liberal Party had passed without dissent in June 2007 a resolution supporting the idea. Interviewed on ABC radio, her brother, Professor Richard Broinowski, said that a likely outcome of proposed Australia–US co-operation on nuclear issues would be pressure for Australia to store American waste. Michael Angwin of the Australian Uranium Association described the fear as "conspiracy theory gone mad," John Howard denied the accusation, and Alexander Downer called it "one of the more wacky theories I've heard." Only time will tell who is right, but don't say you weren't warned!

The University of Melbourne research group admits that estimates of the cost of de-commissioning nuclear power stations vary wildly, from US$300 million in the US to about a billion pounds sterling in the UK — in other words, in Australian dollars, from about $400 million to about six times that much. In Sweden, France and the US, operators have been levied a charge to provide for waste disposal. These funds have been used for research into the problem. France has set aside over A$100 billion for this task, but there is still no credible technical solution.

The House of Representatives committee was also predictably sanguine about the waste issue. It said in its summary of recommendations that the Council of Australian Governments should "examine whether ... there is in fact a potential role for Australia in the back-end of the nuclear fuel cycle" (a euphemism for being a waste dump). The Executive Summary says:

> The volumes of high level waste (HLW) are extremely small, contained and have hitherto been safely managed. The committee finds that HLW has several features which lends itself [sic] to ease of management: very small volumes; the radioactivity is contained in the spent fuel assemblies; it decays at a predictable rate; and is amenable to separation, encapsulation and isolation ...
>
> The generation of electricity from a typical 1000-megawatt nuclear power station ... produces approximately 25–30 tonnes of spent fuel each year. This equates to only three cubic metres of vitrified waste if the spent fuel is reprocessed ...
>
> HLW is accumulating at 12,000 tonnes per year worldwide. The IAEA states that this volume of spent fuel, produced by all the world's nuclear reactors in a year, would fit into a structure the size of a soccer field and 1.5 metres high ...
>
> To date, there has been no practical need and no urgency for the construction of HLW repositories.
>
> There is an international scientific consensus that disposal in

geologic repositories can safely and securely store HLW for the periods of time required for the long-lived waste to decay to background levels. While plans for geologic repositories are now well advanced in several countries, finding sites for repositories has been problematic. This has been due in large part to a lack of public acceptance. "Not in my back yard" arguments about the siting of repositories have been fuelled by misperceptions about the level of risk involved ...

So there you have it; it isn't really a problem, according to the parliamentary committee. A typical power station only generates 25–30 tonnes of waste a year, so that's only about 1000 tonnes over its lifetime. It's only accumulating at the moment at 12,000 tonnes a year, so we don't have to worry. The industry is confident it could be stored safely, if only you ignorant people would realise how safe it is and let them bury it in your backyard. How comforting it is to have reassurances of this kind.

The new line being pushed by the nuclear lobby is that nuclear power stations offer reliable greenhouse-friendly electricity that is potentially cheaper than renewables, while the problems of waste management and weapons proliferation are at least in principle capable of being solved. I hope it is clear from the analysis I have presented that this is wrong in every respect. Nuclear power is neither a timely nor a cost-effective way of reducing greenhouse pollution, waste management remains an unsolved problem, and the difficulty of securing fissile material makes proliferation a more worrying issue today than it was thirty years ago when the Fox Report was published. So, if the facts are so clear, why is it being promoted?

There is no simple answer. In analysing the debate I was reminded of another exercise I undertook twelve years ago. I was commissioned by the commonwealth government of the day to write a report about the population debate, another complex question. Those in favour of population increase included traditional expansionists who believe that growth is inherently good, minority expansionists who wanted to see more people of European background coming here or having children, migrant organisations who sought to protect the rights of their members to sponsor relatives or friends, growth-oriented state or territory governments such as those of Queensland and the Northern Territory, some humanitarians who wanted a more generous approach to refugees, some economic libertarians who see restrictions on the movement of labour as a form of protectionism, and some commercial groups with a direct financial interest in expansion such as the housing industry and estate agents.

Opponents of population growth included some environmentalists who see the level of consumption threatening natural values, urban residents who see growth reducing their quality of life, infrastructure providers worried about their capacity to support rapidly growing urban numbers, some unemployed people who see migrants competing with them for jobs, cultural protectionists who see the Australian way of life being changed by successive waves of migration, unashamed racists who see migrants from some ethnic backgrounds as inferior, some Indigenous people who see further growth as reducing still further their chances of protecting their culture, and so-called "off-shore humanitarians" who think we should be helping disadvantaged people where they are rather than enabling them to come here. So it was no surprise that the debate was often confused, with such a complex mixture of motives and world-views on each side.

Something similar applies to the politics of the nuclear debate. In favour of nuclear power are some obvious vested interests, like the employees of the Australian Nuclear Science and Technology Organisation and those who are retired from that body but still support its general ethos. There are commercial interests: companies that mine uranium or are prospect-ing for it, as well as firms trying to find support for other parts of the nuclear industry like enrichment. But there are also technophiles, people who are by training or inclination strongly attracted to the idea of a tech-nical fix for the energy problem. Thus, for instance, the editor of the house journal of the Australian Academy of Technological Sciences and Engineering contacted me when I was elected a fellow. He asked me to submit an article putting the case against nuclear power in Australia, hav-ing failed to find any of the existing fellows willing to take that stance. More recently, some scientists, usually from the biological fields, have become so alarmed by the possible impact of climate change on natural systems that they are prepared to support nuclear power as the lesser of the two evils. There are probably also some members of the community who have been persuaded by the recent "information" campaign to come to the same view.

Just as those in favour of nuclear power do not form a homogeneous group, their opponents are also diversely motivated. Environmentalists are almost universally hostile to the threat posed to the natural world by the nuclear industry at all stages, from uranium mining to waste and weapons. There are NIMBY groups who do not want the nuclear industry anywhere near their own backyard. There are well-informed physical scientists who do not accept the assurances of the nuclear industry that the problems can be solved. There are economists who can't see any way of making the costs of nuclear power look credible. There are Indigenous groups who resent being pressured to approve mining as a toxic trade-off to obtain their basic rights. There are political pragmatists who think that it is very unlikely to have public support in the foreseeable future. And there are also commercial interests: companies that are developing renewable energy industries and are understandably resentful of the government resources being squandered on the nuclear cause.

So there is no simple or absolute explanation for any particular politician supporting or opposing either uranium mining or nuclear power. For example, I don't think there is one simple reason for John Howard's sudden embrace of the pro-nuclear agenda in mid-2006, but I can see several elements that might have contributed. One is the increasing recognition that the denial of global climate change was politically untenable. After a decade of systematically undermining the logical responses – improved efficiency and renewable energy – the Prime Minister could hardly embrace them after a sudden epiphany, but he could perhaps go for the nuclear option. The political appeal of this choice was the possibility of turning it into a positive, re-badging himself: no longer asleep or in denial about climate change but a "man of steel," prepared to take the hard decisions in the national interest, not afraid to look at controversial alternatives. There is no doubt that the "debate" has been a great distraction for the government, deflecting attention from a shameful decade of denial and obfuscation.

It is also an issue which allowed some possibility of wedging the ALP, which has been divided on the nuclear question ever since the time of the Fox Report. In the 1970s, the fault-line was broadly between those unionists who would support almost anything that provided jobs for their members, and the white-collar members who saw the environmental and social problems of the nuclear industry. After the Fox Report, the latter view prevailed and the ALP took an anti-uranium stance, highly critical of the Fraser–Howard open-slather approach. The "three mines policy" was a bodgy compromise negotiated by the silver bodgie himself, Bob Hawke, to defuse what was seen as an electoral liability for the South Australian ALP, which was required by party policy to oppose the proposed Roxby Downs mine. The Coalition in that state supported the mine and promised it would bring untold wealth, so the local ALP under John Bannon felt it had to neutralise the issue. There was no logical or moral principle behind the notion that one uranium mine was acceptable but others would not be, especially as the one that was supported was likely to become the world's largest. It was the ultimate pragmatic compromise, likened by some to supporting chastity while running the biggest brothel in town.

The policy of Labor opposition to new uranium mines persisted from 1984 until earlier this year. At the time of the Howard announcement, there was a very wide range of views in the ALP on the nuclear issue. Anthony Albanese as environment shadow minister had campaigned vigorously against proposals for new uranium mines. Peter Garrett, both as lead singer for Midnight Oil and as president of the Australian Conservation Foundation, had been trenchantly anti-nuclear before being elected to parliament. At the other extreme, some front-bench members, such as Martin Ferguson, had been working for years to undermine the ALP policy, even joining with the Coalition majority on the House of Representatives Committee to support the outrageously pro-nuclear report discussed elsewhere in this essay. From the outside, it seemed that both Kim Beazley and Kevin Rudd fell for the dishonest line peddled by some elements of the press that support for uranium mining was a key indicator of economic responsibility.

The pragmatists took the ALP a decisive step further to the right at the 2007 national conference, essentially adopting the Coalition policy of unreserved support for uranium mining and export. No logical argument was adduced for the change, with union heavyweight Bill Shorten resort-ing to the pathetic line that it had to be supported so the conference wasn't seen as "rolling the leader in an election year." Even that wasn't enough to persuade a majority of the accredited delegates; a group had to absent themselves and allow proxy-holders to vote for the change. This has cer-tainly blunted the previous ALP message of "renewables, not reactors." While opposed to the development of nuclear power in Australia and to other elements of the nuclear industry, such as uranium enrichment and radioactive waste dumps, the national ALP is quite comfortable about allowing unlimited expansion of mining and export (although the state branches in WA and Queensland do not support new uranium mines).

The government's support for the nuclear industry could also be seen as yet another aspect of the knee-jerk support for George W. Bush which has been a defining characteristic of John Howard's prime ministership. Just as we have uncritically sent troops to ill-considered military adventures proposed by the current US president, we have also been happy to support his blocking of progress toward a global agreement on climate change by refusing to ratify the Kyoto Protocol, and have even been prepared to sell out the national interest in a ludicrous "free trade agreement" that offered very few benefits to Australia. As noted earlier, I have a nagging suspicion that we are being lined up to be the dumping ground for the US's nuclear waste, now that it has proven impossible to inflict it on their own citizens. The Howard government brought in legislation to over-ride the Northern Territory government and impose a low-level radioactive waste dump, having failed in its previous hamfisted attempt to locate the facility in South Australia. Since the Howard government has already signalled its intention to change the legislative framework for nuclear issues, it could conceivably enact the legal basis for the storage of high-level waste.

There would remain the highly contentious issue of siting if the Com-monwealth government tried to impose nuclear reactors on any local community. This might be characterised by a variant on the tired old NIMBY (Not In My Backyard) approach: NIMBE (Not In Marginals

Before Elections). Dr Switkowski, now chairing the board of the Australian Nuclear Science and Technology Organisation (ANSTO), told a recent Adelaide conference that it would require "community education" to develop support for such a proposal. I was immediately concerned that we would not see anything that you or I would call education. Educate, in this instance, is an irregular verb: I educate, you advocate, he brainwashes. Dr Switkowski assured the conference that the material would be "factual," but there are two points to be made about that assurance. One is that there are facts which might appear to support the case for nuclear power, and other facts which show that there is no credible case. Which set of facts do you think a Howard government would collect and distribute?

Just as the Switkowski group appears to have collected those facts that show the nuclear industry in the most positive (or least negative) light, so any campaign of "education" would almost certainly be intentionally slanted. Mr Howard, like any clever politician, has refined the art of making statements that are not factually incorrect but are deliberately misleading, such as the breathtaking claims about interest rates made before the last election. He didn't actually say he would keep interest rates down, because he clearly couldn't. He didn't actually say that he had successfully kept them down when he was treasurer, because he hadn't. So he said that interest rates will always be lower under a Coalition government than an ALP government under the same circumstances, a claim that cannot be tested because we can't have two parallel economies to test the theory, leaving the clear impression that he could be trusted to hold down interest rates.

We can expect any "education" about nuclear power to be carefully crafted misinformation, involving nothing so glaring as factual errors which would allow the material to be discredited, but which nevertheless amount to a misleading package. Take for example, one egregious instance of this sort of misrepresentation by the nuclear lobby. In 1980 Sweden held a referendum after a period of public debate to determine the country's future involvement in nuclear power (a model of public participation we might well follow). Voters were given three choices: to close down the reactors immediately, to allow the existing ones to operate for their design life but then close them and build no more, or to allow the industry to expand. The voting was roughly 40 per cent for immediate closure, 40 per cent for a phasing out of the industry and 20 per cent for its continuation. The Swedish government saw this as 80 per cent of the people voting for the reactors to be shut down, either immediately or gradually, and adopted that position. In Australia, however, the nuclear advocate Leslie Kemeny told a forum that the Swedish people had voted 60–40 in favour of keeping or expanding the nuclear power capacity of the country!

There is also the problem for any information campaign that some "facts" do not exist: they are unknown and in some cases unknowable. In a famous paper about "trans-science," the US nuclear scientist Alvin Weinberg argued that there is a category of questions which can be couched in the language of science and are clearly scientific in their thrust, but which cannot be answered in terms that are acceptable to scientists. He gave as two examples the operating safety of nuclear power stations and the health effects of small doses of radiation.

In the case of power stations, he said, you cannot know how safely they operate over their full life-cycle until you have built and run thousands of them. There will always be a case for caution in dealing with the unknown. When the contraceptive pill was being introduced, a clinician was asked whether it was safe for women to take the pill for decades. He replied that as the answer to this was not known, a responsible stance would be to say that no woman should be on the pill for twenty years until its safety had been shown by thousands of women being on the pill for twenty years! In that case, the assurances of the medical profession were accepted and millions of women went on the pill; it subsequently became apparent that there were health risks associated with taking it, but even in hindsight most women feel the risk was worth the benefit of being able to control their fertility. In other cases, the assurances of science proved to be tragically wrong, as in the view that the prions causing "mad cow disease" could not be a risk to humans. So we could accept the assurances that Chernobyl was an isolated incident, that there is no real risk from hundreds or thousands of nuclear reactors, but we cannot know in advance if those assurances are correct. One concerning factor is that acceptance of nuclear power implicitly presumes eternal peace. Just as power stations and dams were seen as legitimate targets in the Second World War, it would be naive to assume that power stations would not be attacked in any future conflict. While bombing a conventional power station interrupts supply and disrupts a nation's capacity to wage war, bombing a nuclear power station would have appalling consequences, spraying radioactive debris over a wide area.

Weinberg's second example concerned the health effect of low levels of radiation. We know from the US bombing of Japan, the Chernobyl disaster and accidental exposures that large doses of radiation have very serious health effects. The probability of damage to the body is linearly proportional to the dose of radiation: double the exposure and you double the risk. What was unknown when Weinberg wrote was whether this relationship extended to low doses. There were two competing theories. Some scientists believed that the same linear relationship would apply to lower and lower doses, so there is no such thing as a safe dose of radiation, just a steadily decreasing risk. The alternative theory argued for a threshold, some critical dose below which no harm at all would be done. The grounds for believing this theory were that there are very large variations in the so-called background level of radiation in different parts of the world, arising from differences in the geological structure. Cities like Armidale or Aberdeen have much higher levels of natural radiation than cities like Brisbane or Oxford because the underlying rocks are quite different. No consistent health anomalies have ever been shown to arise among people living in those places with higher levels of radiation, giving some comfort to those who would like to believe that radiation in small doses is harmless.

Weinberg pointed out that it would not be morally acceptable to conduct controlled experiments, irradiating different groups of people by varying amounts and observing the health impacts, so we just would not know unless it became possible to gather statistical evidence that might tilt the balance one way or the other. In fact, since then there has been very detailed study of the health impacts of the low levels of radiation from different types of housing in Europe. Scientists are now leaning toward the view that there is no threshold, but the case is far from closed. Weinberg's general analysis still holds: there are some questions which just cannot be answered with a confidence we would like to have before making decisions that potentially expose people to risk. Where the facts are not known, assertions cannot be made responsibly, but those that are made also cannot be shown to be wrong. There is a German saying that no trader calls out "Rotten fish!"; those producing either education or propaganda will naturally tend to show the cause they are supporting in a favourable light. Compare these two statements, either of which could easily appear in a brochure placed in your mailbox soon:

> Nobody can demonstrate that a nuclear power station in your suburb would be a health hazard. Professional agitators will claim that there is a risk, but they can't prove it.

> Nobody can prove that a nuclear power station in your suburb would be safe. We know there will be radiation and it might be a risk to you and your children. The nuclear industry will claim it is safe, but they can't prove it.

Both statements are arguably correct, though I think the first one is deliberately misleading; even if the risk can't be accurately quantified, there is undeniably an additional risk to be considered. There may well be people in the nuclear industry who would object to the second statement. The first is designed to reassure you and the second is intended to alarm you. Which one do you think is more likely to be included if the Prime Minister is educating the community about nuclear power?

The final question about education or propaganda is how effective it is likely to be. When the advocates of the nuclear industry say that they need to educate people about the safety of reactors, they are implying a deficit model. They, being informed about nuclear technology, know it is safe, whereas you, being ignorant, are worried. If only they could pass on to you their superior understanding, you would be as confident as they are of the safety and efficacy of nuclear power, so you would joyfully accede to it. But two examples to do with genetic modification of food suggest that this is an over-simplified view of the world.

Some years ago, the Australian Consumers' Association ran a consensus conference on the question of genetic modification of food. A small number of volunteers were selected to participate, excluding people with pre-existing strong views for or against the technology. They were allowed to choose the experts they wanted to hear and question in an intensive long weekend of discussion. At the end of the process, the group was much better educated about the technicalities of the issues than it is feasible to imagine the overall population ever being. But the group was no more sympathetic to the technology at the end of the process than they were at the beginning. Those who were inclined to be in favour before the process tended to be better informed supporters at the end, while those who were critical at the beginning tended to be better informed critics after the educational experience.

One possible explanation is that deciding whether given benefits, real or alleged, are worth the risks, known or unknown, is essentially a value judgment. Sometimes you and I reach different conclusions because we have different levels of information; if my conclusion is based on wrong information and you correct my misunderstanding, I might change my view. But on other occasions we could have the same information but reach different conclusions about whether the risk is justified, based on different values. For example, I have often heard economists say that consumers are not rational. What they mean is that people make decisions that are not solely based on economics, as when they volunteer to pay extra for "green power" – I pay a levy on my electricity bill, but my lights are no brighter and my computer no safer than they would be if I paid the standard electricity bill, so I am paying more money and not getting any extra service. I do it because I hope that my willingness to contribute will swell the demand for clean energy and change the overall pattern of supply, thus contributing to a cleaner world and slowing climate change. In fact, it is not rational to make decisions solely on the basis of economic issues, ignoring social and environmental questions. But I can understand why other people make different choices.

Similarly, we all have different views about how much tax should be collected and what services should be provided by our governments. Some people make strenuous and even illegal efforts to avoid paying tax, while others regard their tax as their reasonable contribution to a decent society. I would be happy to pay more tax to have better schools, but would be very unhappy to pay more tax to support the development of Australian nuclear weapons; I imagine there are others who have little interest in the quality of public schools but would support increased military spending. While programs of education or propaganda can change the information we have available when we make decisions, they may not change the values that underpin our choices. Suppose our scientific knowledge had advanced to the point where we could accurately predict how many more cancers would result if we went for nuclear power in Australia. For the purpose of this argument, it doesn't matter whether the number would be one extra case a century or a hundred a year. Deciding whether that is acceptable or not is a value judgment.

Different people will legitimately reach their own conclusions because we do not have an agreed view on acceptable risk. We were shocked by the Bali bombing when eighty-eight Australians died, most of them young people. But every three weeks on our roads about 100 Australians die, most of them young people; far from being shocked, we refer to "the road toll," as if this carnage was the appropriate price to pay for the privilege of using roads. So there seems little prospect of there being community agreement about the acceptability of nuclear power, even if we could say accurately what the risk would be. The fact that we can't give any such estimate is a further complication.

In the case of advanced technology, there is a particular problem that could be called the May Effect. Dr Bob May, now Lord May, gave a fascinating paper to the UNESCO World Conference on Science in 1999 looking specifically at the question of attitudes to genetic modification of food. He reminded us that most scientists believe in the deficit model, which maintains that those who are informed understand the benefits and welcome the new technology while the ignorant are needlessly worried. He then looked at international comparisons of attitudes. The highest level of support for genetically modified food was in the US, where the level of scientific literacy is the lowest in the OECD, with about a quarter of all adults believing that the Earth and all species were created in six days only a few thousand years ago and a significant number still thinking that the Sun goes round the Earth! As you move up the scale of increasing scientific literacy, May observed, you get decreasing support for genetic modification, with the opposition strongest in the Scandinavian nations with the world's highest levels of scientific literacy. His explanation for this was that only the ignorant believe that scientific advance is an unmixed blessing. If you understand the science, he said, you will be aware that it is always some sort of Faustian bargain; there are always costs as well as benefits. So, he said, we should not be surprised that those communities which have the highest levels of scientific education are the least likely to believe the bland assurances that all will be well.

This cautionary tale should be a warning to all those who suppose that public hostility to nuclear power will be dissolved by education. Some of the most trenchant opponents of the nuclear push have doctoral qualifications in physics and could not remotely be accused of failing to understand the issues. Scientists such as Dr Alan Roberts, Dr Mark Diesendorf, Professor Jim Falk, Dr Jim Green and myself fall into this category; we have strong backgrounds in physics and know how risky is the course being advocated. We also understand the science of climate change and the urgency of a response; in my case, I have been advocating reduction of our greenhouse pollution for twenty years. But we see nuclear power as a foolish response to the problem.

So how might the issue affect the 2007 election? In one sense, it is still more lead in John Howard's heavily laden saddle-bags. I don't think many people have been impressed by his dramatic endorsement of the nuclear option as an option for slowing global warming. Surveys show overall opposition to nuclear power in Australia, even though it is now common for the question to be worded to increase the chance of a positive response. The Coalition backbench strikes me as being distinctly nervous about the electoral impact of the issue. The panic was palpable when the Australia Institute put out a list of electorates that might be likely sites for nuclear power stations. On the other hand, it is not an issue that hands Kevin Rudd a clear advantage. He is seen as being more credible on the issue of climate change than John Howard, and he has a better overall response to it. But he actively supported the change to the ALP policy on uranium mining, muddying the previously clear message of "renewables, not reactors." The kinder interpretation of this support is that he has implicitly accepted the line that the world needs our uranium. The only other explanation is that he is prepared to support commercial development at the expense of the environment and national security; which would make him a younger and slightly more socially responsible version of John Howard. And there are also clear divisions among his colleagues, evident at the ALP conference when some frontbenchers spoke passionately against the change while others were in favour. The Greens and the Democrats, both strongly and consistently against further Australian involvement in the nuclear industry, have a clear strategic opportunity at the 2007 election. Whoever becomes prime minister is likely to have to do deals with the other major party in the Senate to counter the hostile opposition from the Greens and any remaining Democrats.

THE ALTERNATIVES

We do not need to take the nuclear path. We should set targets for renewable energy in the same way as have progressive nations in the northern hemisphere. We could aim at generating 10 per cent extra electricity from renewables by 2010, 20 per cent by 2015 and 30 per cent by 2020. These are realistic targets based on existing technology.

Nor need they involve massive price increases. As far back as 1992, the Department of Resources and Energy estimated that we could get 30 per cent of our electricity from renewables at no significant extra cost. The technology has improved dramatically since then, despite meagre funding compared with the resources poured into the nuclear option. A recent report by the Australian Conservation Foundation and other environmental NGOs, titled *A Bright Future*, showed that we could still get 25 per cent of our power from a mix of renewables by 2020, despite fifteen years of inaction since the 1992 report. Barry Naughten, formerly a senior economist with the Australian Bureau of Agricultural and Resource Economics (ABARE), summarised the global view as follows:

> a major model-based analysis by the International Energy Agency in June 2006 analysed cost-effectiveness of technologies that could together reduce emissions at 2050 by 60 per cent. Not all these scenarios included expanded nuclear. Indeed, the IEA noted that many of its member-states opposed such expansion. But even in a scenario where such expansion was assumed, nuclear was found to account for only 6 per cent of the total emission abatement compared with 44 per cent from improved end-use energy efficiency, with the remaining 50 per cent from a variety of other technologies.

Hugh Saddler, Mark Diesendorf and Richard Denniss have developed a detailed energy scenario for Australia in *A Clean Energy Future for Australia*, showing that we could dramatically reduce our greenhouse pollution without recourse to nuclear power. It is worth noting that a strategy of this sort would be much better for employment and the economy generally than the present approach. A 2003 Commonwealth report, *National Framework for Energy Efficiency*, estimated that domestic, industrial and commercial energy use could be cut 30 per cent using measures that would repay the initial investment in less than four years. That approach would create more than 10,000 jobs in activities such as retro-fitting buildings, installing solar hot water systems and replacing inefficient equipment, mostly in regional Australia. Efficiency measures and a real commitment to renewable energy would employ about as many people as the entire workforce of the coal industry. So we should treat with appropriate scorn the sort of nonsense the Prime Minister has trotted out about the possibility of a clean energy future: that it would throw thousands out of work, or that it would require replacing every coal-fired power station with an equivalent nuclear reactor. These statements are factually incorrect.

The Prime Minister has even resorted to what Guy Pearse called "arguably the biggest lie in [his] 33-year political career." Blatantly misleading parliament, Howard claimed "a 50 per cent cut in Australian emissions by 2050 would lead to a 10 per cent fall in GDP, a 20 per cent fall in real wages," attributing those figures to ABARE. That government agency has been justifiably criticised for overstating the cost of reducing emissions, but Mr Howard's claims distorted its findings by omitting a crucial phrase: "compared with business as usual." In fact ABARE did not say the economy would shrink and real wages would fall if we cut our emissions; their report projected a 246 per cent *increase* in GDP by 2050, not a 10 per cent fall, and an 81 per cent increase in real wages! Embellishing even more, the Prime Minister went on to claim that there would be "a staggering 600 per cent rise in electricity and gas prices," when ABARE's estimate was 80 per cent, slightly less than the projected growth in real wages. So even the models of ABARE, which could not conceivably be accused of being constructed to undermine the Howard line, show that meeting ambitious reduction targets involves no real economic pain. We would only sacrifice a small fraction of the anticipated increase in our wealth for the sake of keeping the planet habitable. On the other hand, we might well be poorer if we went for the expensive option of nuclear power.

Due to government inaction, it would probably now take until some time in the 2040s to achieve the goal of getting all our electricity from a mix of renewables. That timescale would avoid premature retirement of existing plant. We would not build any more coal-fired power stations, gradually bringing on-line a range of renewable technologies as the old units are retired, at the same time reducing demand by sensible cost-effective measures, with the possibility of some gas plant to meet peak power needs. I would like to see other states follow the lead of South Australia and outlaw electric water heating in favour of solar, heat pumps or gas. We should set a serious target for biofuels in the transport sector as well as requiring cars to be more efficient and investing properly in public transport. Governments at all levels should be modelling best practice in buildings, operations and transport. In mid-2007, the ABC broadcast a TV series called *Carbon Cops*. The first example showed a typical affluent family how they could dramatically reduce their wastage of energy. Efficiency measures and simple lifestyle changes saved them $7000 a year in fuel and electricity bills, so that they could easily afford to spend less than a tenth of that on green energy and carbon offsets. They finished up being carbon-neutral instead of being serious polluters and more than $100 a week better off! It was a graphic example of the adage that saving the planet need not cost the Earth. We could live at the same level of material affluence using much less energy, and causing much less greenhouse pollution, without resorting to nuclear power.

Above all else, we should set a long-term target to cut our greenhouse pollution by 2050 and take it seriously. Our past approach, having demanded at Kyoto the world's most generous target and now making no serious effort to cut emissions, is a source of shame to thinking Australians. The only reason it can be claimed that we are roughly on track to meet our Kyoto target is that our target was uniquely generous. Not only did the Australian delegation obtain permission to increase pollution when most other industrialised countries accepted reduction targets, it also successfully demanded the addition to the Kyoto Protocol of what is known around the world as the "Australia clause." This gives us credit for stopping broad-scale land clearing, which inflated our 1990 baseline by about 30 per cent. If the rest of the world had not given us that special treatment, our greenhouse pollution rate in 2010 would be 40 per cent above our Kyoto baseline. Few countries have so recklessly increased their release of greenhouse gases. If I accept the "Australia clause" and re-set our 2010 figure to 10 per cent above the 1990 baseline, a linear reduction to a responsible 2050 goal gives interim targets of 15 per cent below 1990 levels by 2020, 40 per cent by 2030, 65 per cent by 2040 and 90 per cent by 2050. Other analysts have suggested a simpler target would be an annual percentage reduction, e.g. 5 per cent less each year until we reach our agreed eventual goal. The point is that we must have a target and a comprehensive strategy for achieving it. In his recent book, *Greenhouse Solutions with Sustainable Energy*, Mark Diesendorf writes:

> Fortunately, we already have sustainable energy technologies that are capable of achieving deep cuts in Australia's greenhouse gas emissions of 50–60 per cent by 2040–60. For stationary energy [uses other than transport], these technologies are the myriad of products and measures comprising efficient energy use, together with solar hot water, bio-energy from crop residues, wind power and, as a transitional fuel, gas (both natural gas and coal seam methane). More expensive renewable energy technologies, such as bio-energy from dedicated crops, solar heat at 100–300 0C and solar electricity, could be further developed to achieve total emission reductions of 80 per cent or more in the second half of the 21st century. If climate change continues to accelerate, these technologies could be implemented earlier.

Diesendorf includes in his book an example of a mix of technologies that could replace a 1000-megawatt coal-fired power station: essentially 375 megawatts of wind power, 206 megawatts of bio-electricity and 220 megawatts of combined-cycle gas power. This approach would reduce carbon emissions by nearly 5 million tonnes a year. It is possible to replace a coal-fired power station entirely by a renewable supply like wind power, but the capacity has to be scaled up to give the equivalent average output over a year. Diesendorf estimates that 2700 megawatts of wind turbines has the same average power output as a 1000-megawatt coal-fired power station, allowing the power station to be retired, as has been done in Denmark. Even with this extra capacity, wind power is still cost-effective. In the UK a 2005 study found the average cost per kilowatt-hour of electricity from wind turbines to be 8 Australian cents, compared with about 15 cents for nuclear power.

In *A Clean Energy Future for Australia*, demand management measures such as solar hot water and improved efficiency are used to reduce electricity demand in 2040 to 14 per cent below the 2001 value. This is a crucial point. Studies that presume continuing growth in energy use tend to reach the conclusion that new renewable capacity cannot be built fast enough. In the medium-efficiency scenario, carbon emissions from electricity are cut by 78 per cent with a supply mix of gas (30%), bio-energy from crop residues (28%), wind (20%), coal (9%), hydro (7%) and solar for afternoon peak demand (5%). This example shows it is entirely realistic to meet Australia's future energy needs with a mix of clean supply options, completely eliminating the need to go nuclear.

A DANGEROUS DISTRACTION

Just as we now recognise that the dangers of asbestos outweigh its benefits, I believe we should also acknowledge that the risks of mining and exporting uranium are not justified by the benefits. I suspect the true motive of some who have called for a debate about nuclear power is to soften us up for expansion of uranium mining and export. Prime Minister Malcolm Fraser claimed in 1977 that "an energy-starved world" needed our uranium. This was a transparent attempt to portray a crass commercial operation as a moral virtue by claiming that the world needed nuclear power. I had a sense of déjà vu when I saw the title chosen by the rabidly pro-nuclear House of Representatives committee for its report: *Australia's uranium – Greenhouse friendly fuel for an energy hungry world*. And the committee put forward an updated version of Fraser's rhetoric: "As a matter of energy justice, Australia should not deny countries who wish to use nuclear power in a responsible manner the benefits from doing so. Neither should Australia refuse to export its uranium to assist in addressing the global energy imbalance and the disparity in living standards associated with this global inequity."

According to these politicians, it is our moral duty to let the poorest people in the world have dirty and expensive energy that will leave them the enduring legacy of nuclear waste. If we were serious about helping the developing nations to have the energy services we take for granted, we would be promoting Australian solar and biomass technology, which is both much more appropriate to their needs and much more likely to provide jobs and economic benefits than expanding uranium exports. We should also curb our own profligate energy use. Despite the hype, uranium accounts for about 1 per cent of our mineral exports, ranking with such metals as tin and tantalum. The most optimistic forecast of the potential annual revenue from uranium sales to China by 2020 is about a third of our current income from exporting cheese. Given that the safeguards agreements have more holes than a Swiss cheese and radioactive waste is more unsavoury than an old gorgonzola, I would be much happier if we concentrated on our cheese sales. Since every gram of uranium exported increases the problem of radioactive waste and increases the amount of fissile material that could be diverted to weapons or dirty bombs, we should be phasing out the industry rather than contemplating any expansion of it.

Deciding whether the risks of any activity justify the benefits is always a value judgment, so I can understand how people I respect have come to different conclusions. The only reason we are even considering nuclear power is global climate change, which demands that we dramatically reduce our greenhouse pollution. This is the most important issue for our time. If we really were in a situation where use of nuclear power was the only way to prevent dangerous climate change, the choice would be very difficult. It might be possible to solve the technical problems of nuclear power and the economic costs would be tolerable if that were necessary to stop catastrophic climate change, but the social and political problems do not appear to have solutions. Fortunately, we do not face the terrible choice between runaway climate change or a nuclear future. From local studies to the International Energy Agency 2006 report, there is ample evidence that we can curb our greenhouse emissions without making a Faustian bargain with the nuclear industry. The scales are weighted very heavily against nuclear power as a response to global warming. It is too expensive, too risky, too slow and makes too little difference. The clean, green path is better environmentally, economically and socially; efficiency and clean energy supply will take us toward a sustainable future, whereas nuclear power would be a decisive step in the wrong direction. The high-cost, high-risk option of nuclear reactors is a dangerous distraction from the urgent task of reducing greenhouse pollution. The 1970s slogan remains true today: if nuclear power is the answer, it must have been a pretty stupid question!

SOURCES

2   "The Fox Report": Ranger Uranium Environmental Inquiry (1976), First Report, AGPS Canberra, pp.5–6.

3   "two fundamental problems": Ranger Inquiry, pp.185, 187.

4   "a small group in the UK": see "When PR goes nuclear", *New Statesman*, 27 May 2005. Available at <http://www.energybulletin.net/6427.html>.

5   ABC Radio, *AM*, 7 June 2006.

5   ABC TV, *Lateline*, 8 June 2006.

5   "The subsequent report": Commonwealth of Australia (December 2006), *Uranium Mining, Processing and Nuclear Energy – Opportunities for Australia?*, Report to the Prime Minister by the Uranium Mining, Processing and Nuclear Energy Review Taskforce.

8   "We have known about the problems of peak oil and climate change for decades": see K.S. Deffeges (2001) *Hubbert's Peak*, Princeton University Press, Princeton, reviewing Hubbert's work since his seminal paper in 1956, and I. Lowe (1989), *Living in the Greenhouse*, Scribe Books, Newham.

8   UNEP (1999), *Global Environmental Outlook 2000*, Earthscan, London. See <http://www.unep.org/geo2000/>.

8   "the idea of peak oil": M.K. Hubbert (1956), "Nuclear Energy and the Fossil Fuels", *American Petroleum Institute Drilling and Production Practice, Proceedings of Spring Meeting*, San Antonio, pp.7–25.

9   "thirty years ago": I. Lowe (1977), "Energy Options for Australia", *Social Alternatives* 1: pp.63–69.

10   "a post-industrial economy": see Barry Jones (1982), *Sleepers, Wake!*, Oxford University Press, Melbourne.

11   "one of the most unequal": see I. Lowe (2005), *A Big Fix: Radical solutions for Australia's environmental crisis*, Black Inc., Melbourne. For more detailed analysis, see K. Norris and I. McLean (1999), "Changes in earnings inequality 1975–1998", *Australian Bulletin of Labour* 25(1), pp.23–31.

12   "a recent study": see United Nations Department for Disarmament Affairs (2002), *Disarmament Studies Series No. 31: The Relationship between Disarmament and Development in the Current International Context*, United Nations, New York.

14   I. Lowe (1989), *Living in the Greenhouse*, Scribe Books, Newham.

14   For a review of the distortions and mistakes in presenting climate science in *The Great Global Warming Swindle*, see <www.csiro.au/resources/pfxg.html>.

16   "the Earth is now warmer": see I. Lowe (2005), *Living in the Hothouse*, Scribe Books, Melbourne.

16   Millennium Assessment Report: see <www.millenniumassessment.org>.

16   J. Lovelock (2006), *The Revenge of Gaia*, Allen Lane, Melbourne.

16   "the world as a whole must reduce greenhouse gas emissions": see IPCC (2007), *Fourth Assessment Report: Climate Change 2007*. For more information, see: <www.ipcc.ch>.

17   "annual public subsidy": see M. Diesendorf (2007), *Greenhouse Solutions with Sustainable Energy*, UNSW Press, Sydney, pp.290–291, Table 14.1.

17   "cutting energy use": see <www.eceee.org> for case studies and a general review.

17   K. Hargroves & M.H. Smith (eds) (2005), *The Natural Advantage of Nations*, CSIRO Publishing, Melbourne.

19   G. Pearse (2007), *High and Dry*, Penguin Books, Melbourne, p.90.

20   "average annual rate of increase": see C. Flavin & J.L. Sawin (2004), "The 'tipping point'", *Renewable Energy World*, May–June 2004. Available at: <http://jxj.base10.ws/magsandj/rew/2004_03/tipping_point.html>.

20   The assertion was made in an editorial in the *Australian*, 27 July 2007.

20   "far too slow a response": see UMPNER Report.

21   "the Fox Report warned": see Ranger Report, pp.185–187.

21   "Mohamed ElBaradei": see <radio.un.org/story.asp?NewsID=2038>.

22   For the scale of waste produced, see UMPNER Report.

22   S. Kim & I. Woods (2004), *Nuclear Fuel Cycle Position Paper*, AMP Capital Investors: <http://www.ampcapital.com.au/corporatecentre/research/sriposition.asp>.

23   "total life-cycle analysis": see <http://www.nuclearinfo.net/Nuclearpower/SSRebuttalResp>.

24   House of Representatives Committee on Industry and Resources (2006), *Australia's uranium – Greenhouse friendly fuel for an energy hungry world*, Parliament of the Commonwealth of Australia, Canberra. Available at <http://www.aph.gov.au/house/committee/isr/uranium/report.htm>.

24   F. Hoyle (1979), *Energy or Extinction*, Heinemann, London.

25   World Energy Council (2005): see <www.worldenergy.org>.

25   G. Cravens (2007), *Power to Save the World*, Alfred A. Knopf, New York, p.374.

27   "it is possible for all people to live": see E. Weizsacher & A. Lovins (1997), *Factor Four*, Allen & Unwin, Sydney.

27   For scale of use of renewable energy, see M. Diesendorf (2007), *Greenhouse Solutions with Sustainable Energy*, University of New South Wales Press, Sydney.

28   For China's energy plans, see <www.un.org/events/wssd/statements/china>.

29   "State government of Queensland": see *Weekend Australian*, 9–10 July 2007, p.12.

31   UMPNER Report, Appendix 1.

36   UMPNER Report, Section 4.4.

37   For discussion of subsidies of aluminium industry, see H. Turton (2002), *The Aluminium Smelting Industry: Structure, market power, subsidies and greenhouse gas emissions,* Discussion Paper 44, The Australia Institute, Canberra. Available at <http://www.tai.org.au>.

38   "research group at the University of Melbourne": see <http://www.nuclear-info.net/>.

39   "a recent Adelaide conference": M. Hibbs (2007), "The Nuclear Renaissance on the Global and Local Scale", paper presented to The International Expert Workshop: Uranium: Energy, Security, Environment, Adelaide, 7–8 June 2007.

40   The Australia Institute: see <http://www.tai.org.au/documents/downloads/MR237.pdf>.

41   For a review of the issue of Australia's possible development of nuclear weapons, see R. Broinowski (2003), *Fact or Fission? The Truth about Australia's Nuclear Ambitions,* Scribe Publications, Melbourne.

43   Dr Switkowski's speech is accessible at <www.pmc.gov.au/umpner/docs/draft_report/launch_speech.doc>.

45   "we should look at uranium enrichment": ABC Radio, *The World Today,* 7 June 2006.

45   House of Representatives Committee Report, p.628.

46   ABC TV, *7.30 Report,* 15 June 2007.

47   "Carlson said": "Uranium to Russia", *Herald Sun,* 17 June 2007. See also <http://www.dfat.gov.au/media/releases/department/d010_07.html>, as John Carlson claimed in this media release to have been misquoted.

47   "Peter Garrett": see <www.petergarrett.com.au/c.asp?id=377>.

47   Dr Clarence Hardy: ABC TV, *7.30 Report,* 15 June 2007.

48   "make the rubble bounce" is still being said in the US Senate. See <http://www.time.com/time/magazine/article/0,9171,922832-1,00.html>.

49   "mini-nukes": J. Sterngold (2003) "Senate debates ban on small warheads", *San Francisco Chronicle,* 21 May 2003. Available at <http://www.commondreams.org/headlines03/0521-10.htm>.

49   Richard Broinowski: House of Representatives Committee Report, p.382.

51   Richard Broinowski: House of Representatives Committee Report, p.380.

51   Ian Hore-Lacy: House of Representatives Committee Report, p.374.

52   ASNO: House of Representatives Committee Report, p.383.

52   Mohamed ElBaradei: statement to 2005 NPT review conference. See <http://www.iaea.org/NewsCenter/Statements/2005/ebsp2005n006.html>.

52   ElBaradei: House of Representatives Committee Report, p.390.

53   C. Keaney (ed.) (2006), *An Illusion of Protection*, Australian Conservation Foundation and Medical Association for the Prevention of War, Melbourne.

53   Weapons of Mass Destruction Commission (2006), *Final Report*. See <http://www.wmdcommission.org>.

54   Safeguards: UMPNER Report, pp.110–111.

55   "safeguards have been diluted": see Kearney (ed), pp.31–32.

55   "sold to Taiwan": "China 'comfortable' with Australian–Taiwan nuclear tie", *Financial Times*, 4 April 2006.

56   "Pakistani minister": ABC TV, *Lateline*, 26 July 2007.

56   "foiled a plot": *Lateline*, 11 June 2002.

56   Gordon Corera: *Lateline*, 9 July 2007.

57   S. Harris (2005), *The End of Faith*, W.W. Norton, New York.

57   T. Findlay (2007), "Implications of the Nuclear Revival for Global Nuclear Governance", paper presented to the International Expert Workshop: Uranium: Energy, Security, Environment, Adelaide, 7–8 June 2007.

58   F. Barnaby (2007), *Too Hot to Handle? The Future of Civil Nuclear Power*, Oxford Research Group, Oxford.

60   ASTEC (1984), *Australia's Role in the Nuclear Fuel Cycle*, Australian Government Publishing Service, Canberra.

61   Helen Caldicott and Michael Angwin, ABC TV, *Difference of Opinion*, 19 July 2007.

61   Helen Caldicott, "Too much haste to nuclear waste", *Courier-Mail*, 8 August 2007.

61   John Howard: "Australia won't become nuclear waste dump: PM", *News.com.au*, 20 July 2007. See <http://www.news.com.au/story/0,23599,22105687-5005080,00.html>.

61   Alexander Downer: See <www.foreignminister.gov.au/transcripts/2007/070720_doorstop.html>.

62   University of Melbourne research group: <http://www.nuclearinfo.net>.

62   House of Representatives Committee Report, pp.lii–liii.

64   "population debate": I. Lowe (1996), *Understanding Australia's Population Debate*, Bureau of Immigration, Multicultural and Population Research, Canberra.

69   Z. Switkowski (2007), "From Nuclear Review to Public Policy", paper presented to the International Expert Workshop: Uranium: Energy, Security, Environment, Adelaide, 7–8 June 2007.

70   A. Weinberg (1972), "Science and Trans-science", *Minerva* 10, pp.209–222.

73   "consensus conference": The Australian Museum (1999), "First Australian

Consensus Conference: Gene Technology in the Food Chain". Available at <www.austmus.gov.au/pdf/layreport.pdf>.

75   R. May (1999), "The Scientific Approach to Complex Systems", paper presented to the UNESCO World Conference on Science, Budapest, June 1999.

76   "likely sites": "Nuclear Power Plants", Australia Institute media release, 27 February 2007. See <http://www.tai.org.au/documents/downloads/MR237.pdf>

77   M. Stevens (1992), "Renewable Electricity for Australia", NERDDC Discussion Paper No. 2, Department of Resources and Energy, Canberra.

77   J. Rutovitz, M. Wakeham and M. Richter (2007), *A Bright Future: 25% Renewable Energy for Australia by 2020*, a report by the Australian Conservation Foundation, Greenpeace Australia Pacific and Climate Action Network Australia. Available at <http://www.acfonline.org.au>.

77   B. Naughten (2007), "Sabotaging Kyoto: Howard's real agenda on climate Change", *Dissent* 23, Autumn–Winter 2007, pp.38–39. See also: International Energy Agency (2006), *Energy Technology Perspectives – Scenarios & Strategies to 2050*, IEA, Paris.

77   H. Saddler, M. Diesendorf & R. Denniss, (2004), *A Clean Energy Future for Australia*, Clean Energy Future Group. Available at <http://wwf.org.au/publications/clean_energy_.future_.report.pdf>.

78   *National Framework for Energy Efficiency*: see <www.nfee.gov.au>.

78   G. Pearse (2007), *High and Dry*, Penguin Books, Melbourne, pp.377–378.

80   M. Diesendorf (2007), *Greenhouse Solutions with Sustainable Energy*, UNSW Press, Sydney, p.341 and Appendix A, pp.346–348.

ACKNOWLEDGMENTS

Patricia Kelly, Peter Christoff, Don Henry and other Australian Conservation Foundation staff kindly read a draft of this essay and made very helpful comments. I would also like to thank Chris Feik, the editor of *Quarterly Essay*, for helpful advice, suggested sources and insightful editing that significantly improved the text.

Correspondence

Philip Ruddock

Apparently Australians are "easily persuaded," "habituated," "desensitised," and we "have only the patchiest record of becoming passionate about great abstractions." The Australian government is "savage," with "dark motives," and we are all a bit obsessed with "security, morality, respectability [and] order."

These are some of the comments in the latest *Quarterly Essay*, authored by Fairfax journalist David Marr.

It would be fair to say that the author is a critic of the Commonwealth government. He says we are "practical," and means it as an insult.

I take a more optimistic view of Australia. I think Australia should be kept safe from the threat of terrorism, and should not be a soft touch for people smugglers. I do not consider that violent street protesters are fighting for the "great abstractions" in life. I think the police do a good job in the main.

Australia is one of the world's oldest democracies. People have been voting and participating in public debates here for far longer than they have in most other countries. The amount of popular participation leading up to Federation, for example, was unprecedented (this includes participation by people who did not then have the vote).

I do not take the view that Australians should be derided as unthoughtful or unsophisticated. Frankly, I would seriously question whether the elites who do take that view are really in touch with the Australian community or only circulate in a cloistered mutual-appreciation society.

Many of the allegations in the *Quarterly Essay* are factually incorrect. Even simple matters, for example the allegations that there are no Commonwealth protections for journalists or whistleblowers, are off the mark. According to the *Quarterly Essay*, "Canberra has none of Washington's provisions to protect whistleblowers who go to the press" and there are "no shield laws to protect journalists."

In fact, there are protections.

Public servants who wish to make a disclosure may do so either within or outside their department. The Public Service Commissioner, the Merit Protection Commissioner and the Commonwealth Ombudsman are all independent bodies to whom public servants may make disclosures. It is against the law for them to be discriminated against or victimised for doing so.

Similarly, the Commonwealth now has shield laws to protect journalists. To say otherwise is just wrong. The author may have been confused about a recent case in which two journalists were prosecuted in Victoria; because of the Victorian law at the time, there was no protection. It would hardly be fair to blame the Commonwealth for the state of Victorian law.

The *Quarterly Essay*'s general thesis is that in comparison with previous administrations, the current Australian government is making information harder to obtain and bullies anyone who dares to disagree. Inaccuracies about journalists and whistleblowers are used in support of this thesis.

There is also an argument that the Commonwealth's immigration policies are part of a broader campaign against journalists. The Christmas Island detention centre was established, apparently, "to keep the cost of investigating these stories very high. It's a long way to send reporters." The truth of course is that the Christmas Island detention centre sends a very strong message to people smugglers and their clients. The message is that there is no fast-track to the mainland. There is no way to jump the queue.

The *Quarterly Essay* also attacks the Commonwealth's sedition laws. Again, the complaint is that they are part of a broader anti-journalist conspiracy. Sedition, it will be remembered, involves inciting resistance to lawful authority, and may include the overthrow of a constitutional order. It is particularly relevant in the fight against terrorism. Terrorists see our democratic freedoms and institutions as a source of weakness. I refuse to let that be the case. Our way of life needs to be protected.

The *Quarterly Essay* accuses the government of bullying journalists. I do not accept that assertion and would suggest that the Howard government has distinguished itself from Labor in this area. When former prime minister Paul Keating did some bullying of his own recently, Marr went on television to describe it as "marvellous ... it was marvellous."

There are legendary tales of the former prime minister ringing journalists to give them a less than favourable character assessment. Keating was also notorious for declining to attend question time. His predecessor, Bob Hawke, was renowned for his colourful response to critics, and there are allegations that the current opposition leader's office regularly berates editors and has issued threats to reporters to not report stories. But the *Quarterly Essay* does not address these.

This is not to say the current government measures itself by the behaviour of the former Labor government. It does not. The former government had lost touch with the electorate by focusing on the "great abstractions" and ignoring the basics. The current government on the other hand, is derided for taking a more "practical" approach.

In his criticism of the general public's lack of dissent, Marr seems to celebrate the violent protests around WTO discussions and the G20 meetings, while deriding the actions of police and governments.

Yet another area of contradiction comes with the decision-making process. When public servants or decision-makers come up with a result he likes, they "turn out to be far less spooked" than the government. At least, this was Marr's take when a series of videos was not banned under classification laws. The reality was the decision-makers were simply applying classification law as they saw it, but when the government sought a review of that decision, for some reason that was an unacceptable act. Even Marr should acknowledge that governments, like the rest of the community, have the right to appeal where that mechanism exists. Where there is uncertainty or doubt, it is a responsible action to follow through.

The essay is riddled with hyperbole. In 2005 Australia apparently did away with habeas corpus, allowing "arrest without charge and detention without trial." According to Marr "it was difficult to see what bedrock rights remained." If Marr was indeed right, in the intervening period there should have been hundreds – no, thousands – of people locked up under this oppressive regime. The facts are that under strict controls, people suspected of having knowledge of terrorist events can be questioned and/or detained for defined (short) periods. Some of these activities need court endorsement or are supervised by judicial officers. There is little resemblance between Mr Marr's hyperbole and reality.

The government welcomes strong debate. It is a cornerstone of our democracy. We should be challenged on our ideas and policies and they should be scrutinised by the electorate so judgment can be passed.

In the final analysis, there is no conspiracy. Papers like the *Quarterly Essay* are presented and critique the government, the High Court has recognised a freedom of political communication in the constitution, and voters have the final say.

Australia is a wonderful country. I think most people are happy enough to recognise that. Others should consider it.

Philip Ruddock

Julian Burnside

David Marr's essay was at once dazzling and depressing. By bringing together a list of events, principally from the past five months, he gives tangible shape to an uneasy feeling – the feeling that some very basic principles are being dismantled before our eyes; the feeling that John Howard's rhetoric is an elaborate set of stage props which have kept most of us distracted and calm while the theatre burns around us.

Howard has stifled and channelled the public debate to such an extent that the boy who speaks of the Emperor's new clothes will not be heard by many, and those who do hear will turn away in denial or despair. Marr's snapshot of the recent life of Australia is almost too distressing to read. Its calm look at what is going on, and how those events are tolerated, ignored or not noticed, tells several powerful stories.

First, that we are losing (if we have not already lost) some basic features of what used to be valued in this country. Those who hold dissenting views are vilified or marginalised, while the crooners in the press fan-club are lionised. The War on Terror leads us to invade Iraq, thus increasing the risk of a terrorist attack, but at the same time democratic freedoms are dismantled in order to reduce the risk again. A peace activist is bundled out of the country for reasons which cannot be revealed – perhaps because Howard wanted to be seen as a war prime minister and talk of peace is somehow offensive or subversive. ASIO and the Australian Federal Police are given unimaginably vast powers, but somehow their bungling does not inspire a sense of greater safety. In short, all that matters is the way it is sold to us, the spin. Drawing on the lessons of history, Howard knows that the key to power is to make sure your message is the only one noticed by the average punter. No matter that the message is a lie to conceal earlier lies; what matters is that it is the dominant message. Provided others cannot be heard or will not be believed if they tell the truth, you can get away with murder.

The second thing we are losing is that deep primal concern about a fair go. Its content has been leached away until all that remains is comforting words of self-justification. We pay lip-service to the idea, and persuade ourselves that uttering the mantra is as good as achieving the thing itself. Superficially, the sudden reversal of public opinion about David Hicks looked like a triumph for the spirit of human rights and a fair go. But of course it wasn't. Much more likely most Australians had written Hicks off as a dangerous fool who deserved about five years for what he did (or might have done, and who cares which?). With five years up, it was time to bring him home. No visible public sentiment about the colossal unfairness of the trial he was about to undergo; no general protest when Ruddock announced that he thought it would be a fair trial, despite the use of coerced evidence and hearsay. And not a squeak about the fact that the fair-go kids – Howard, Ruddock and Downer – had abandoned Hicks to his fate over the previous five years. And even now, no public alarm that Mamdouh Habib was rendered to Egypt for "enhanced interrogation" for the benefit of America and with the knowledge of the Australian government. Where was our capacity for courage when these things were revealed to us?

Would Australians of the 1930s or the 1960s have tolerated a government so careless of one of its sons, whatever his faults? Would past generations of Australians have tolerated the barefaced lying, which is now so much Howard's trademark that occasional truth seems an accidental embarrassment?

We do not trust Howard or his government. Trust went years ago. A range of lapses, from the GST to the exposure of the children overboard fable, made sure of that. Then the lies that took us into Iraq stripped away even the possibility of trusting him. But when did we lose the capacity to be outraged by it all? That's the great loss. When did the majority of Australians settle back into their modest homes of increasingly astronomical price and decide that we can't stop him lying to us, we can't stop the hypocrisy, so who cares?

Because Marr's essay is principally concerned with the past five months, it does not reach back to Al-Kateb's case – that scar on the legal landscape of this country. Al-Kateb had come to Australia as a boat person. He applied for asylum and waited out his time in Woomera. His claim to asylum was rejected. Rather than stay longer in detention during an appeal, he asked to be removed from Australia. The *Migration Act* provides that a non-citizen without a visa must be detained until he gets a visa or is removed from Australia. Al Kateb waited a long time in Woomera: removing him was complicated by the fact that he is stateless, and there was no country which could be forced to take him back. The anomaly faced by Al-Kateb and two or three others could have been fixed by a simple amendment of the *Migration Act*. Instead, the Howard government argued all the way to the High Court that Al-Kateb – innocent of any offence, not posing any threat to the community – could be held in detention for the rest of his life. On 6 August 2004, the High Court held, by a majority of 4 to 3, that the government was correct.

The wickedness of the argument still takes the breath away. It is still a shock to recall that the government was willing to contend for such a result. Al-Kateb's case ought to be branded on the conscience of every Australian. 6 August should be declared a National Day of Remembrance, when we recall that a government once argued for the lifetime detention of an innocent man, simply because he asked for help and we couldn't force someone else to take him off our hands. But Australians do not know about Al-Kateb's case, because the media largely ignored it. None of the Howard cheer-squad – Andrew Bolt, Piers Akerman, Miranda Devine, Janet Albrechtsen – expressed even mild dismay that the government could contend for such a result. Between them they have a reach into most Australian homes. But so muted was their response to Al-Kateb (if they responded at all) that the wickedness of the government's position remains, effectively, hidden. They are condemned by their silence. And they think, like Howard, that we don't need a Bill of Rights because we have the majestic protections of the common law. Try telling that to Mr Al-Kateb.

It breaks the heart and tests the spirit to see the reckless cruelty of this government, and the way its image has been polished and propped up by the Howard fan-club in the media. Patronage and preference are valuable and flattering assets. It takes Spartan honesty to resist them, and Howard's collaborators in the press are not made of such stern stuff. David Marr, and a few others, have the wisdom to see what is going on, and the honesty to report it. If Australia ever comes to its senses again, Marr will be among the heroes while Howard's media acolytes will be consigned to the footnotes.

Julian Burnside

HIS MASTER'S VOICE	*Correspondence*

Peter Shergold

I have recently read two articles, written a generation apart and from opposite ends of the world, but reflecting on the same Westminster tradition.

The first was David Marr's *Quarterly Essay*. With customary eloquence, it mourns an Australian public service cowed by the Prime Minister into abject fear and supine silence. "His master's voice," according to Marr, has "neutered Canberra's mandarins."

By way of evidence, Marr presents what he perceives to be the unacceptable intimidation of courageous whistleblowers. He gives five instances of leaking, of which he attributes four to public servants: the theft of six cabinet submissions on indigenous issues; the provision to the *Herald Sun* of a draft government policy proposal from the Department of Veterans' Affairs (which, Marr notes, "had been jettisoned before the paper hit the streets"); an internal report to Customs on the failings in airport security; and a private record of a discussion between the Australian foreign minister and the New Zealand high commissioner.

The passion of the language – which imagines John Howard "thrashing public servants" and establishing "political control" – glosses over the most fundamental of facts: in none of these four examples was the transmission to the media or the Opposition of confidential government documents driven by the need to expose behaviours that were unlawful, unethical or even unacceptable. There is no suggestion that corruption, nepotism, misappropriation or inappropriate conduct were exposed. In each case the motivation – to which Marr is naturally sympathetic – was that the leaker thought that either the policy decisions or their public administration did not meet their own perception of the national interest.

They were not exposing a breach of the Australian Public Service (APS) Code of Conduct. Quite the opposite: the *Public Service Act*, passed with bipartisan support in 1999, makes it clear that such behaviour is itself a breach of the Code.

Setting the standard for good governance, the Code requires that an APS employee must maintain appropriate confidentiality about dealings with ministers and must not make improper use of inside information.

The behaviour Marr admires is also at odds with section 70 of the *Crimes Act*. It is an offence for current or former Commonwealth officers to publish or communicate Commonwealth information where there is a duty not to disclose. This is not some recent measure designed to silence dissent. Non-disclosure offences have been a feature of the Act since 1914.

Let me be blunt. The individuals cast in an heroic light by Marr, however well-intentioned, did wrong. That is why some of them ended up facing disciplinary proceedings in the APS or being cross-examined in a courtroom.

That does not mean that there is not a role for whistleblowers in the APS. There is – in law. Ironically (from Marr's perspective) it was under the Howard government that the *Public Service Act* for the first time recognised and provided protection to whistleblowers who seek to report suspected misconduct in the public interest. An agency is not allowed to victimise or discriminate against employees because they have alleged a breach of the APS Code of Conduct. If whistleblowers are dissatisfied with the agency's response, they can refer their grievance to the Australian Public Service Commission, which has statutory independence.

They do so. Each year complaints are received, for example about people who are alleged to have falsified information or engaged in bullying or harassment. I know, from personal experience, of a whistleblower who reported a manager for wrongly claiming travel allowances. The manager was dismissed from the APS. I know of others removed from the APS for transmitting pornography.

What Marr doesn't accept is that public servants do not have a moral responsibility or enjoy legal protection for leaking documents simply because they believe the government has got it wrong. Good governance depends upon senior public servants being able to provide ministers with frank and fearless policy advice in confidence. If they cannot do so, ministers – concerned that policy briefs or discussions will be made public by disgruntled public servants – may be tempted to look for advice only from the advisers they can trust. That would serve to undermine the relationship between government and public service, which is the very foundation of the Westminster tradition. It is a recipe for poor public administration. By sidelining public servants, it would politicise decision-making.

This is not a popular position. I thought Marr's essay was hard-hitting until I chanced upon Mike Carlton's June 30–July 1 piece in the *Sydney Morning Herald*.

Carlton makes Marr seem mealy-mouthed. My view that public servants should be responsive to the government of the day in serving the public interest would, it seems, have allowed me to send people to death camps and gas chambers. My "paper-clipped view of the world" is "facile bluster." "Hitler, Stalin and Pol Pot would have applauded that one. Auschwitz? None of your business, dear boy, the government knows what it's doing. The Gulag Archipelago? The Khmer Rouge Year Zero? Not your moral responsibility, old chap."

The humour is more offensive than incisive. Believing that Customs should have taken more account of your report on airport security is not quite a matter of Year Zero proportions. Thinking that the government should have been more generous in its proposed increase in funding for veterans is rather different from considering whether to transport them to the Gulag. No Australian public servant can imagine that they are facing such moral challenges.

Of course, some public servants on some occasions may feel that they are so opposed to a particular government public direction that, as professionals, they would prefer not to be involved in that area. In my career I've helped people transfer to areas of the public service in which they felt more comfortable. It's entirely appropriate.

Similarly, and particularly when I was the CEO of the Aboriginal and Torres Strait Islander Commission (ATSIC), I had long discussions with members of my staff who found the demands of professional public service too constraining. For indigenous people, committed to delivering for their community, the need to work to government can prove too heavy a burden. In some cases, following soul-searching, they decided that they could achieve more working in a non-partisan manner within the system. Others, with my full support, decided to move into the community sector, where they could advocate publicly. Both were ethical positions. Leaking is not.

The second article, and in truth the far more intellectually challenging one, was a convocation address given at the University of Leeds in 1978. A thoughtful presentation about the British civil service, it was published in the *University of Leeds Review* (vol. 21, 1978) as "Power in Government – A Chinese Puzzle." It was given by a political adviser, namely Jack Straw, who had been president of Leeds Students' Union in 1967–68 (when, along with his future government colleague John Prescott, I was finishing my undergraduate degree at the University of Hull). By 1978 Mr Straw had already been political adviser to two ministers (Barbara Castle and Peter Shore) and was, at the time of his remarks, Labour candidate for Blackburn. Of course, he went on to become home secretary and secretary of state for the Foreign and Commonwealth Office in the government of Tony Blair and the leader of the House of Commons. He has now been appointed by Gordon Brown as the new justice minister and lord chancellor.

The ABC on Sunday evenings recently ran a BBC production entitled *Life on Mars*. The story – and I followed it closely – concerned a 2005 police detective, knocked into a coma in a car accident, who regains consciousness in Manchester in 1973. He faces culture shock as he comes face to face with misogyny, racism, violence and corruption, which, apparently, he hadn't encountered in the modern police force. This is, after all, a work of fiction.

Reading Jack Straw's address, one is persuaded that going back to the 1970s would probably involve an equal jolt for today's civil servant. In 2007 the views of David Marr are common currency. The public service is politicised. Its leadership is allegedly subservient to the will of its political masters. The electoral and ideological priorities of party politics are imposed by governments on their public officials. The bureaucrats are manacled. Such are the perceptions portrayed in the media.

Live at Leeds in 1978, the popular concern was the exact opposite: the fear was of bureaucrats reining in ministers. Straw's address was prompted by a stereotype that "at its upper levels, the civil service is undermining the very foundation of our democracy, by manipulating ministers as if they were puppets, and by ignoring or subverting the will of a democratically elected House of Commons." The senior civil servants, opined the *News of the World*, were "the people who really rule Britain."

Such views were not confined to Whitehall or Fleet Street. Indeed, they were reflected almost exactly in the Antipodes. Gough Whitlam at the federal level and Don Dunstan at the state level were equally concerned at the perceived capacity of senior public servants to subvert political will to their greater experience and conservatism. According to Dunstan, his senior public servants "almost without exception … set out to mould ministers' views to their own and to manipulate the minister to prevent marked change."

Straw's address has the shock of the old to a public servant who now accepts as holy writ the *Public Service Act, 1999*. An ideal before which I genuflect daily – that the APS must perform its functions in an impartial manner – Straw dismisses as misperceived mythology. He accepts that public servants must be impartial in executing and administering government policy without corruption or favour, through the fair application of transparent rules. But when it comes to providing expert policy advice, the "Service is in fact expected to be highly partial to the government of the day."

Straw's point is that public servants necessarily have their own views on public policy, for why else would they have been attracted to work in government? There is nothing wrong with that. Indeed, it is those public servants who openly acknowledge and argue for their cause who best advise ministers, "rather than the grey men who pretend no views and who display an attitude to politics little different from that of a monk's to sex." But – here's the rub – those well-formed opinions need to be tempered by what we now, somewhat cautiously, describe as responsiveness, but which Straw, far more assertively, identifies as loyalty.

This, from the vantage point of 2007, sounds provocative indeed. It is bad enough, from Marr's perspective, that public servants are responsive to the direction of the prime minister (as, incidentally, is explicitly required under the *Public Service Act*). That they should aspire to be loyal would almost certainly be deemed to be even more dangerous.

Loyalty carries the connotation of being attracted to another by affection or devotion; of faithfulness, inspired by fond feelings of allegiance. In Marr's eyes, I am pretty confident, loyalty would be an inappropriate quality on which to build the relationship between the prime minister and his secretary. At first blush, the concept seems at odds with the apolitical independence on which the Australian Public Service prides itself, and which our critics believe we have forsaken. Shackled by contract appointments and induced by the lure of performance pay, we are judged to have sold our souls to executive authority. It is, from this jaundiced perspective, a loyalty misplaced and undeserved.

I beg to differ. Straw's point – and it is convincingly argued – is that governmental power is highly intricate and diffuse. In this complex milieu, public servants should be open about their views, should present them forcefully, but should always be partial to the decisions of the government of the day. "What seems to be central to the relationship [between ministers and officials] is not some fanciful concept of impartiality," Straw told his audience, "but whether officials are willing loyally to serve a government even where its views are very different from their own."

Loyalty, as I interpret it in this context, is not subservience. It is devotion to the notion of representative and responsible government. It is faithfulness to the institutional framework of democratic governance. It has the character which Mark Twain ascribed to it in *A Connecticut Yankee in King Arthur's Court*: "the citizen who thinks he sees the commonwealth's political clothes are worn out, and yet holds his peace and does not agitate for a new suit, is disloyal: he is a traitor."

Loyalty, from this perspective, is not unthinking fealty or begrudging silence. It is not symptomatic of the corruption of public debate or of the undermining of standards of governance. Rather, it is the willingness of a departmental secretary to argue her or his opinion frankly and fearlessly; to do so confidentially; and, when a decision is made (as it is appropriately made) by government, to accept it and to execute it with determination. It is loyalty to the challenging discipline of a public service in thrall only to the principles of democratic governance. It is partiality not to political party or to a particular prime minister but to successive governments, each of which will be served with equal dedication. It is the loyalty of public service, born not of weakness but of strength of conviction.

Peter Shergold

Joan Staples

David Marr spells out a list of attacks on public debate – some crude, some more subtly honed – by the Howard government. His essay is important, for it shows the coherence of the attacks and "the ugly pattern they reveal."

As Marr points out, Howard has continued to express rhetorical support for "parliament, the courts and 'a strong free press,'" despite having "worked to curtail them almost from the day he took office." Unfortunately, this rhetoric is not matched by Howard's treatment of civil society. Its democratic role has been systematically attacked by a coherent, many-faceted campaign, stretching over eleven years and reinforced by the Institute of Public Affairs.

A strong, dynamic civil society, encompassing community groups and NGOs, can reflect back to us the richness and diversity of our aspirations, our needs and our culture. It can generate competing ideas, alternative possibilities and creative policy options. It is ever-changing, reflecting changes in society's values and priorities, and it can be filled with contradictions and competing views. A vibrant civil society and a vigorous public debate are measures of a healthy democracy. Unfortunately, in Australia today, the Howard government appears to be promoting a model of democracy that would remove any advocacy role for NGOs.

Since its election in 1996, the federal government, assisted by the Institute of Public Affairs, has campaigned to stop NGO advocacy. It has used many means, but the aim has been consistent – to promote a model of democracy in which NGOs do not contribute to public debate. The government's approach is consistent with neo-liberal public choice theory, and Howard's language often parallels that of public choice theorists. NGOs are accused of interfering with the market, or of representing "special interests" and of being "unaccountable." Consistent with the government's policy of smaller government, NGOs are filling gaps left by the withdrawal of government services. These NGOs are congratulated if they do good works, such as planting trees or feeding the homeless, but pilloried if they comment on public policy. It is a very narrow view of representative democracy in which elected representatives are seen as the only legitimate and accountable participants in public debate.

The list of measures imposed since 1996 is long. De-funding is only one of them, although it has received the most publicity. The NGO sector was reeling from funding cuts from the moment the Howard government was elected, and those affected represented some of the poorest and most disempowered Australians. By 2002, 50 per cent of peak groups within the social-services area had lost significant amounts of funding and 20 per cent had lost funding altogether, largely because of their public advocacy. Every year this process continues across all parts of the sector.

Forced amalgamations have been used to silence alternative views, and purchaser–provider contracts ensure that NGOs deliver services according to the government's agenda rather than their own. By far the most insidious mechanism used has been confidentiality clauses. These appear almost universally in funding contracts and forbid organisations from commenting publicly without the approval of the minister or the department. They are effective, efficient silencing mechanisms and turn NGOs into government contractors.

Organisations that have tried to work without government funding, or to increase their proportion of private funding, have found themselves losing long-held tax-deductibility status if they undertake public advocacy. Since 2003, the government has pursued various paths to this end. First, they drafted the *Charities Bill 2003*, which was dropped just prior to the 2004 election after a concerted campaign, particularly from the churches. However, in 2005, two ATO Rulings relating to public advocacy and tax-deductibility status were successfully implemented. As a result, a number of groups are no longer able to offer tax-deductibility to donors because of their advocacy role.

A recent case has the whole NGO sector agog. AID/WATCH is a very small NGO; it monitors how Australian overseas aid is spent, encouraging efficiency and transparency. It was informed in October 2006 that the Tax Commissioner had withdrawn its tax-deductibility status. Its request for an internal review was recently rejected. The most worrying development is that the ATO review is far more draconian than the 2005 ATO rulings, which were intended to set guidelines. AID/WATCH is now appealing to the Administrative Appeals Tribunal. If the precedent of the AID/WATCH ruling is applied to most Australian advocacy organisations, any public advocacy activity could lead to the revocation of tax-deductibility status, even if that advocacy is only one small aspect of a charitable group's activities. Currently, the Public Interest Advocacy Centre (PIAC) in

Sydney has also lost its status and has sought an internal review from the ATO. The AID/WATCH result does not auger well for PIAC.

But there's more. The *Electoral and Referendum Amendment (Electoral Integrity and Other Measures) Act 2006* requires NGOs to report to the federal government on aspects of their activities. This has no relationship to elections and has echoes of the calls by the Institute of Public Affairs to license NGO activity. This is the same Act that closed the electoral rolls early and removed prisoners' right to vote.

Since 1999, in parallel with the government's efforts, the Institute of Public Affairs has been conducting a strong public campaign to undermine NGO legitimacy. The pervasiveness of their material against NGOs in the general media is significant.

How has this onslaught happened without the general public becoming aware? Certainly, it is disappointing that more NGOs have not drawn attention to the issue. It seems to have crept up, undermined and demoralised them before they were aware that it was happening. The enormous diversity of the sector, which is a positive for democratic debate, has also worked against them, as there are few forums where the whole NGO sector comes together. Loss of institutional knowledge is exacerbated by the use of volunteers and the high turnover of personnel because of low wages and stressful working conditions. Lack of resources means that few organisations can afford to do the research that might have shown them what was happening; committed workers focus on the urgent day-to-day issues. With no Bill of Rights, I believe our society can only have a limited civic dialogue on defending such rights. As well, apart from the Council for Civil Liberties, we have very few NGOs whose specific interest is to promote democracy. We have assumed that we are operating under a democratic model, and that our democracy does not need defending. Unfortunately, we were wrong.

One can only imagine what further mechanisms a re-elected Howard government might come up with to stop NGO advocacy. Would a conservative Labor government under Kevin Rudd dismantle the confidentiality clauses, ATO rulings, restrictive legislation, etc.? Our democracy is weaker because of this decade-long campaign against NGOs, and it is not clear how soon this damage will be remedied.

Joan Staples

John Hartigan

David Marr's essay weaves such a tapestry of alleged lies, deception, censorship, intimidation and persecution that, if we believed it all, Australians should be in a state of despair.

While I agree with Marr some of the time, I can't accept much of his reasoning. Debate in Australia is vibrant and intense at all levels of society and through all media: newspapers, radio, television, at public meetings, through the internet, and in journals like this.

The problem I see is the degree to which the flow of information that generates or fuels informed debate has been stifled. When information is suppressed, our right to know how we are governed and how our courts dispense justice is diminished. Our democracy loses some of its spark.

Unlike Marr, I think there are many underlying causes and I am optimistic the problem can be fixed.

Marr's passionate analysis of life during John Howard's eleven years as prime minister is undermined by the zeal and doggedness of his ideology and jaundiced by his dislike of the man.

The problems we now face have occurred at the hands of Australian governments of all political stripes and at federal, state and local levels. Many hundreds of statutes, some federal, some unique to different states, have cumulatively created a wall of prohibitions which hamper what Australians can know about how our governments and courts function. It is, quite frankly, unhelpful to lay all the blame at the gates of the Lodge.

Some of the worst examples of the erosion of free speech can be seen in the adoption of "spin" at all levels of government and business. Debates on issues as important as this should be conducted with a view to achieving change, rather than polarising positions so that problems simply become entrenched.

Government decision-makers are unlikely to be swayed by rhetoric describing

Howard as an evil object of derision. For example, Marr's statement that "after being belittled for most of his political career, Howard came to power determined public debate would be conducted on his terms." Belittled for most of his political career? Really? Only by his political opponents.

Marr applauds actor Terry Serio's "devastating" portrait of Howard in the stage-show *Keating! The Musical* that made him "a figure of fun, but strangely unfunny." Do these observations advance the cause of free speech? Marr accuses the government of discrediting its critics to undermine their arguments. Isn't Marr guilty of the same?

Howard may well have come to power determined that public debate would be conducted on his terms, but show me the politician who doesn't. There is no doubt he has deliberately built a public-relations machine that ensures the "correct" spin is applied to stories affecting his government. It rivals the propaganda machines of previous governments.

But it is disingenuous to suggest that the erosion of free speech has come about as a result of a Machiavellian blueprint carefully implemented over just the past decade and by just one man.

The erosion has been gradual, over at least three decades, and has occurred at the hands of Commonwealth and state governments of all colours. Still, an event that happened halfway through Howard's tenure significantly compounded the problem – the September 11 terrorist attacks. Marr contends that September 11 was just one of a string of events, including the development of the internet, that "changed everything." I believe that September 11 differed immensely because it was an attack on democracy and capitalism and on innocent human life that until then was inconceivable.

The September 11 attack created the climate of public acceptance for introduction of strong measures to counter the terrorist threat, and this was heightened in Australia by the Bali bombing and our participation in the "coalition of the willing" in Iraq.

These events led to a string of anti-terrorism laws that give rise to intrusive surveillance, holding of suspects without charge and curbs on the reporting of security matters by journalists. While the government regards this as a practical approach to extraordinary events, and the public generally sees it as a necessary evil, there must be balance between security and preservation of civil liberties and the public's right to know. The recent Haneef saga is proof enough that even in times of heightened security, there must be an open process.

If citizens are to participate effectively in a democracy, form opinions freely and protect their rights and interests, they need access to information directly, or via the media on their behalf. But across all levels of government, this balance has shifted away from the people to governments, which makes today's freedom of information laws unworkable.

The incidents are numerous. Just recently in New South Wales, despite repeated attempts, access was denied to an Education Department report on violence in schools. We were also not permitted to know which pubs have the highest levels of alcohol-related assault and robbery. These surely are things that the public should be allowed to know.

At the Commonwealth level, News Ltd is still smarting from the costly two-year battle between the *Australian* and the treasurer, Peter Costello, for the release of details of the effect on taxpayers of bracket creep, and of the first home owners' scheme. Costello believed release of this information was "not in the public interest." The High Court agreed, but I believe the media's role is to lift the veil on exactly this kind of information.

Over the twenty-five years during which the Commonwealth's *Freedom of Information Act* has been in place, decisions like this have chipped away at the integrity of the Act. An entrenched culture of resistance to disclosure of information has developed and technological changes render it at odds with the way the modern media operate. It's time for a wholesale overhaul of the Act.

Debate is stifled at other levels too, as with the recent conviction of *Herald-Sun* reporters Gerard McManus and Michael Harvey for refusing to divulge the identity of someone who embarrassed the government by leaking information about the workings of veterans' affairs policy.

It has become commonplace for the Federal Police to investigate journalists to identify leaks, and to pursue relentlessly public servants suspected of being informants, even when the information they have leaked is patently in the public interest. The man charged in the Harvey–McManus case was convicted and later freed because of lack of evidence, but he lost his job. This could be perceived as deliberate intimidation to demonstrate the consequences for any other public servant who might consider "leaking".

Perhaps the worst such case followed the *Australian*'s disclosure of lax security and organised crime at Sydney airport, which was found by an inquiry to be chillingly accurate. But rather than fix the problems, the government unleashed the Federal Police to seek and destroy the whistleblower.

It's my view that in a healthy democracy there would be no need for whistleblowers because governments would be transparent when it came to matters of genuine public interest. Unfortunately, there are times when governments get things very wrong and exposure is necessary. The security issues at Sydney airport were serious and exposure of them led to an inquiry and a $200-million upgrade. Was the decision taken by the whistleblower to expose the problems, or by the *Australian* to publish them, in the public interest or not?

Across all Australian jurisdictions, there must be a process – and protection – for public servants to make public interest disclosures. But given that even with sound protection, some public servants will not use the process, it needs to be accompanied by laws that allow journalists to protect the identity of their sources in cases of public interest.

Ruddock's recent tinkering with the *Evidence Act* to give judges the discretion to decide whether to force journalists to give up their sources is inadequate. The new Act does not provide real protection for journalists; the burden of proof remains on the journalist to show why they should *not* be compelled to reveal their source. It should be the other way around: the prosecution should be required to show why disclosure is necessary.

Judges must also take some responsibility for the lack of transparency. An important issue overlooked by Marr is the propensity for judges in all juris-dictions to close access to courts and suppress details of cases, often with scant reason.

Our media are buckling under more than a thousand court suppression orders preventing publication of certain facts from court cases. Some of these, for example protecting the identity of an undercover police operative, are clearly justified, but many are not. For example, is it fair that a public figure may be protected from embarrassment by having his identity in a court case suppressed? And should an entire anti-terrorism trial be closed even though not all the infor-mation presented is related to national security?

It seems our courts increasingly view the media as a nuisance. No doubt we are sometimes, but shoving us away and denying us access to the workings of our justice system is dangerously shortsighted. Democracy relies on the fact that justice is not only done, but is seen to be done.

Recently Australia's defamation laws (previously a mish-mash) were made uniform. While not perfect, the defamation laws have improved vastly, and this leads me to be much more optimistic than David Marr. The significant progress made shows how, with leadership at the Commonwealth level, improvement and consistency could also be achieved in areas such as suppression orders.

So how did the erosion of public debate happen? Marr believes it happened because Howard in 1996 set out on a deliberate campaign to cow his critics, intimidate the ABC, gag scientists, silence non-government organisations by threatening their finances, neuter Canberra's mandarins, curtail parliamentary scrutiny, censor the arts, ban books, criminalise protest and prosecute whistle-blowers. I'm less paranoid.

I also have trouble accepting Marr's analysis of the Australian character. He believes Australians project themselves as easygoing larrikins with contempt for authority, when in reality they passively accept it.

He traces this to the mood of the British settlers from whom most of us descend. He says those who settled America did so to secure freedom in a time of repression; hence their preoccupation with freedom. Meanwhile, those who settled Australia were content with British law and customs and compliance with authority. But then he makes the extraordinary claim that Australian children are taught not to speak. "It's a big part of our upbringing, learning to shut up, to listen, to wait until we're spoken to," he says. "Somehow the habit of holding back has been drilled into the character of the nation."

He continues: "Perhaps at some obscure level we still think keeping quiet will do us good when Canberra tells us what we can say, what we can know, when we can speak." I grew up in a different Australia. The one I see encourages children to think and talk and develop self-confidence and be part of a vibrant, open, multicultural and prosperous society.

And the evidence of this is everywhere. In Australia we talk, we question, we read, we listen to dissenting views and we work for change. We're doing it now: three months ago, an unprecedented coalition of Australia's major media organisations formed to work for improvements to free speech. I'm proud that News Ltd is part of that coalition and I'm confident that we really can effect change.

Of course, with freedom comes responsibility and we must continually strive to ensure that our media deserve to represent the public in their right to know. I accept that we haven't always been as careful and responsible as we should be in our reporting.

But occasional errors by the media should not allow us to lose sight of the far bigger issues and freedoms at stake. We should all accept that a healthy democracy is also a place where people argue, disagree, criticise and speak out fearlessly when they believe it's important to do so.

John Hartigan

Tom Switzer

Travel across the country and ask Australians about the state of the nation, and the response is almost universal: we're a free, prosperous and self-confident people, at ease with ourselves. Which makes it very strange that some fellow citizens – and in particular those who think they represent the nation's conscience – spend much of their time fretting and wailing about what Robert Manne calls "the increasingly authoritarian trajectory of the political culture" under John Howard.

From *Silencing Dissent* to *The War on Democracy* to *Do Not Disturb: Is the Media Failing Australia?*, the message is the same. In the words of Clive Hamilton and Sarah Maddison, the Howard government, in cahoots with a "right-wing syndicate" of media commentators, has "systematically targeted independent, critical and dissenting voices" in order "to ensure that its values are the only values heard in public debate."

David Marr's *Quarterly Essay* is the latest we-are-silenced screed, though admittedly his thesis is more sophisticated than the aforementioned publications. Whatever one thinks about his politics, it is hard to deny that Marr is one of the nation's most eloquent writers and commentators. He also makes some plausible points.

Marr is right, for instance, to raise doubts about unfettered power being granted to the executive or its security agencies, and about the new barriers to reporting in Canberra. This is especially the case given the trend towards some government secrecy, with laws to punish whistleblowers and the prosecution (and conviction) of journalists who refuse to reveal their sources.

All true. But in attributing virtually any act of dishonesty and suppression in Australian public life to John Howard, Marr seriously overstates his case. It is one thing to argue about impediments to free reporting. It is another thing to claim that critics of the government – dissenting voices – are being systematically silenced by that government, and that Australia, moreover, is heading down an autocratic path. Marr is muddying the waters; the two things are very different.

In his treatment of the national psychology, Marr accepts, with hardly any reservations, his literary hero Patrick White's thesis that Australians are philistine and stunted. But however true that was in the 1950s, it surely does not serve as any kind of description of today's Australia.

Fifty years ago, the cultural landscape was indeed as flat and unvaried as the proverbial Australian sheep station. Newspapers were mostly dull and parochial. There were no opinion pages, no radio talkback programs and obviously no websites or blogs. Although Robert Menzies' conservatives were in power, a shallow, reflexive, progressive orthodoxy prevailed. This was a time when the leading historian, Manning Clark, went to the Soviet Union and wrote a glowing book called *Meeting Soviet Man*, in which he likened the ideals of Vladimir Lenin to those of Jesus Christ. And so it went.

Today, however, the cultural landscape is much less isolated and more diverse. Marr laments: "The exaltation of the average is back in a big way." This at a time when society offers unparalleled opportunities for an increasingly aspirational class of Australians. Marr warns that Howard has "cowed his critics" and "muffled the press." This at a time when the marketplace of ideas has never been so crowded.

True, Marr concedes that "commentators fill opinion pages arguing the opposite" to the government. But that doesn't stop him from concluding that "the steady constriction of public debate under Howard has aroused no deep concern in Australia." Never mind the many dissenting and anti-government columnists such as Phillip Adams, Kenneth Davidson and Alan Ramsey. And never mind publishers such as Scribe and Black Inc. and magazines such as *Dissent*, *Arena*, *The Monthly* and *New Matilda* online – all of which publish overwhelmingly critical, at times hostile, items about the Howard government. The *Age's* opinion page, incidentally, does not even host a weekly conservative columnist.

The point here is not to suggest that the Left still controls the public debate, but that a broad cross-section of Australians is engaged in a healthy conversation about the issues of the day great and small. And this debate is taking place not just in the opinion pages, magazines and books, but also on radio, cable television and, increasingly, in blogs and internet chat rooms. In the Howard era, technology has lowered the barriers to publishing. The result: virtually everyone has a seat at the debating table.

As for the claim that Howard muffles the press, how does Marr account for the fact that the *Australian*, the *bête noire* of metropolitan sophisticates, often embarrasses

Canberra with exclusive stories such as the AWB and children-overboard scandals as well as, most recently, the Mohamed Haneef barrister leak that damaged the Australian Federal Police's case against the terror suspect?

And how does he account for the fact that Howard, unlike his predecessors, has never threatened legal action, much less violence, against journalists? The same, alas, can't be said about his Labor opponent who has tried to censor or kill controversial newspaper analysis of his life and times. For example, when a *Sun-Herald* reporter planned to publish a story that directly contradicted Kevin Rudd's account of his upbringing, Rudd's media adviser privately warned her: "We'll have 100 people ready to roll tomorrow morning to trash you and your paper."

Howard's approach to the media is more dignified and, dare one say it, more democratic. When he thinks he has been misrepresented in the press, he just corrects the record on the run. When his government is challenged over some administrative or public policy embarrassment, he will go on talkback radio or hold a news conference to field questions from the public and from journalists. Unlike his immediate predecessor, moreover, he never calls journalists at some ungodly hour to berate them about their work, and he virtually never misses parliamentary question time. All of this is a far cry from the authoritarian and unaccountable leader who "is almost superhumanly reluctant to engage in frank debate."

There are other problems with the Marr thesis. For example, he implies that any government funding-cuts to the arts hurt democracy. Yet isn't democracy more likely to be threatened when tax-dollars are subsidising plays and films of only one particular ideological slant? He suggests the Victorian police were being undemocratic when they sought to arrest violent protestors at the G20 summit. Yet was it democratic for the demonstrators to terrorise shoppers in downtown Melbourne and hurl barricades at law-enforcement officers?

In Marr's telling, Joe Hockey's use of parliamentary privilege to "monster" a self-described independent industrial-relations specialist amounts to a suppression of free speech. It's more accurate to say that the minister merely highlighted David Peetz's pro-union biases and connections. If an allegedly independent expert appeared as an expert witness in a court case, any barrister worth his pay would subject him to thorough cross-examination on his perceived biases, and any judge would take such biases into account. But when an "independent" expert is subjected to similar scrutiny in the political arena, Marr deems such behaviour to be evidence of the government's "brutal" and "ruthless" agenda. In any case, Peetz still freely speaks and writes about the perils of WorkChoices.

Marr attacks the "profoundly undemocratic" Howard for overturning the Northern Territory's euthanasia laws because "Australians endorse euthanasia overwhelmingly." Marr is wrong. The law was overturned by the Federal Parliament in a conscience vote in which there was no declared government or opposition position. That aside, there is something intellectually dishonest about his argument. For example, if the death penalty were reintroduced – and a majority of Australians say they support capital punishment – would Marr be so fond of people-power? Surely, what's good for the goose is good for the gander?

Herein lies the great weakness of the Marr essay. Any government decision with which he disagrees is interpreted as a threat to democracy. Yet the many government decisions he detests – citizenship tests, border protection, anti-terror laws – reflect the true democratic will of the country. They also reflect the political and cultural realignment of the nation during the past decade. Whereas once conservative ideas were considered outside the boundaries of serious (and morally respectable) consideration, today they represent the political mainstream. Marr derides the term mainstream, but on the great battlefields of history, economics, welfare, education, citizenship, indigenous politics, national sovereignty and values generally, conservative ideas and those of classical liberalism increasingly, although not completely, prevail.

Now, Marr has every right to despair at this trend and he has every right to challenge the public-policy pronouncements of the government. But it is wrong to equate this cultural sea-change with a drift towards authoritarianism. It is, in fact, a drift towards democracy. Howard, after all, has won four elections in a row. So, when Marr, Hamilton, Manne and co. warn that the government has stifled "independent" thinkers, silenced debate and corrupted democracy, that's code for saying: we no longer set the political and cultural agenda. That is the real message of his *Quarterly Essay*.

Tom Switzer

Waleed Aly

In an address to the Young Liberals' national conference in Hobart in January 2005, soon after the Coalition's 2004 election victory, health minister Tony Abbott argued that John Howard had successfully nurtured a team of "new conservatives." This was Howard's "rehabilitation of conservatism in Australia," the result being a conservatism that manages to "define itself in terms of what it's for, not just what it's against."

Gone were the old conservatives, with their "seemingly perpetual grumpiness," who did not construct "conservative utopias." For Abbott, Howard had managed to craft a positive, even relaxed and comfortable, conservative vision; to "create a new story for Australia … as a force for good in our own right."

Abbott's language is moderate. This new conservatism, as he describes it, is far from objectionable. But within it, perhaps, are subtle clues to the philosophy behind the governmental behaviour to which David Marr so vehemently objects. Marr alleges that for its own ideological ends, the Howard government has behaved with a pointedly undemocratic belligerence, and that its conduct erodes the principles and conventions on which Australian public debate is based.

Certainly, it is difficult to dispute that Howard has changed the face of Australian conservatism during his time in power. Classical conservatism valued education and cultural refinement; this government and aligned media commentators regularly disparage artists and intellectuals as an irrelevant and undesirable elite out of touch with middle Australia – as though it is their job to represent public opinion. With its emphasis on individual rights, conservatism once tended towards libertarianism – as reflected in Menzies' decision to name his party Liberal. The values of a society grew organically from below, from the individuals who constituted it. This government instead seeks to articulate them from above, hence its voluminous Australian values rhetoric. The classical liberal ideal of the small state has been replaced by an increasingly intrusive polity, particularly in the realm of defence and counter-terrorism. Even that famous conservative pragmatism has gone missing at crucial junctures: our participation in a war to bring freedom and democracy to the Middle East has a positively progressivist resonance.

Where are all the classical conservatives? It seems they are not in government – at least not vocally. The Howard government looks instead to have borrowed heavily from neo-conservatism – where moralism meets governmental power, sometimes in illiberal ways. The hallmarks of neo-conservatism are present: the belief in a strong, activist state, the alliances with big-business interests, the nostalgic celebration of family values and the threat to promote them through censorship (for example, during last year's *Big Brother* controversy), and the promotion of mass culture (in this case via an Australian values discourse) rather than the acceptance of chaotic and divergent cultural currents. These properties are what distinguish neo-conservatism from its classical cousin, and they appear to be central to Marr's objections.

Howard loves to talk of people being "mugged by reality." It's one of his favourite phrases. Instructively, perhaps symbolically, it is borrowed not from classical conservatism, but from Irving Kristol – the seminal ideologue in neo-conservative thought. Kristol considers the touchstones of neoconservatism to be patriotism, a strong military and an expansionist foreign policy. For the moment, that more closely describes Washington than Canberra, but the differences are rapidly shrinking.

As it happens, I consider Marr's case to be overstated. Several of the episodes he adduces – like his account of the brutalisation of G20 protestors – while disturbing, seem to say more about the police than about the Howard government. On the other hand, it is also possible to caricature Marr's position – and many of his critics do. In my reading, he does not charge the Howard government with gagging debate so much as frustrating it by arguing in belligerent ways that often do not engage the issue at hand. People may still shout their dissent in Howard's Australia – indeed, Marr is an example of this – but they risk being unfairly assaulted in the process or swamped by a torrent of name-calling.

The most vivid example Marr presents is that of Professor David Peetz, who found himself accused of moral equivocation on terrorism after releasing a potentially embarrassing report on the effect of the government's industrial relations legislation. The remarkably bellicose, deceitful, rhetorical onslaught from government MPs was startling: workplace relations minister Joe Hockey accused Peetz of believing that "the President of the United States … is evil" while "good is on the side of the terrorist." This has no relationship whatsoever to industrial relations. As it happens, it has no relationship to truth, either. Marr informs us that the allegation derives from a poem Peetz composed, and the words describe the perspective of a character within it. At no point does Peetz validate this character's view.

Such is the way of neo-conservative politics. It matters not that the allegations against Peetz are nonsense. Here, "truth" is created by relentless declaration until a consensus of submission emerges. What matters is certainty. George W. Bush famously maligned his 2004 election opponent John Kerry by joking that Kerry could debate himself endlessly. What Bush offered – and what Howard offers via his sustained attacks on straw men like moral relativism – is certainty in an insecure world.

Neo-conservative moralism often descends into an assertion of absolutist symbols – what Marr calls the "lazy, brutal assertion of power at the expense of public debate." The ruler is henceforth impervious to interrogation. In this newly moralised landscape, those who would dissent need not be engaged with; they can simply be dismissed. To put it extremely, they are evil. In this way debate, while not formally gagged, can be short-circuited. And in our context, terrorism is the most potent symbol of all. Once it was applied to Peetz, the argument was over, and Peetz was no longer good media talent. The symbol terminates discussion. A new political correctness has emerged.

Which brings us to former terrorism suspect Mohamed Haneef. The basic facts can be recounted briefly. Haneef's cousins have been implicated in failed terrorist attacks in Britain. He had given his SIM card to one of them. Federal police, erroneously believing the SIM card was found at the scene of the crime, handed the brief to the Commonwealth prosecutor who charged Haneef with recklessly – not knowingly – supporting terrorism. From an early stage, this looked like guilt by association. The case was so weak it ultimately fell apart. Even before it did, a Brisbane magistrate granted Haneef bail. That is extraordinary for a terrorism case: bail is only granted if "exceptional circumstances" can be shown. It is difficult to imagine how a magistrate could have reached this conclusion unless the weakness of the case was obvious. Eventually the charges were dropped.

The conduct of the federal government is instructive. Within hours of Haneef being granted bail, immigration minister Kevin Andrews decided to cancel Haneef's visa on the grounds that he now failed the character test under the *Migration Act*. Haneef, insisted Andrews, had an association with people involved in criminal activity. What was so sinister about this association, Andrews wasn't saying. Clearly Haneef was related to terror suspects and had communicated with them. Is that sufficient? Certainly the government's lawyers attempted to argue that a visa could be cancelled on the basis of almost any association: "a cup of coffee, a picnic with the kids." If Haneef's association with his cousins was so sinister, why the need for such a broad interpretation?

The timing of Andrews' decision was extraordinary, giving the distinct impression that it was a hastily cobbled attempt to keep Haneef detained, as well as to cast doubt over him. For nearly two weeks, the information on which Andrews based his decision was kept confidential. When finally Andrews did release information, he did so selectively and out of context. We heard of snippets from emails between Haneef and his cousin suggesting that Haneef leave Australia immediately, using his new-born daughter in India as an excuse. Nothing in the information so far released is direct. What there is, according to Haneef's lawyer, was put to Haneef in a police interview. Whatever Haneef's responses to them, they clearly were insufficiently implicating to support criminal charges. At the time of writing, the transcript of the relevant police interview – the one that might provide some context – has not been released to the public.

With the charges dropped after a gruelling month, Haneef was released from prison. He left almost immediately for India to be with his family, especially the new daughter he had not yet met. It was then that Andrews provided his most outrageous commentary, spinning Haneef's rapid departure as a sign of guilt, as something that justified, and even heightened, Andrews' suspicions. In truth, the greater challenge was to imagine any reason Haneef would have had to stay in Australia. This was the country that had detained him without charge for an inordinate period, kept him in solitary confinement when he was charged on the basis of a flimsy case, then cancelled his visa and promised to deport him. Suspicious behaviour was thin on the ground. Only a man determined to launch an *ad hominem* smear could have made Andrews' comment. Neo-conservative truth-by-declaration returns.

Clearly this approach has worked for the government in the past. Marr's provocative suggestion is that our ingrained national apathy is responsible for this; our trusting of authority, contrary to our popular myth. This may well be true; unfortunately, Marr doesn't develop this idea with sufficient rigour, beyond anecdotal assertion, to judge. Yet, as Marr notes, *ad hominem* belligerence, while far from new in Australian politics, has grown more savage in Howard's Australia. This suggests a philosophical shift is at work in Australian politics that is not simply rooted in traditional Australian acquiescence to authority. It seems to me we are witnessing something of a new political phase, even if it does echo the past.

Tony Abbott was right. Australian conservatism has changed. The question is whether the political landscape has changed irrevocably with it.

Waleed Aly

Gerard Henderson

So David Marr has advised *Quarterly Essay* readers that I am "pompous." Oh dear.

In *His Master's Voice*, David Marr wrote: "When I quizzed Henderson, he admitted he hadn't seen the DVDs and said he didn't think he had to." The reference is to Sheik Feiz Mohammed's DVD – known as *The Death Series* – an extract from which was shown on Channel 9 News on Sunday, 15 April 2007.

This is a considerable embellishment. When David Marr quizzed me about this matter at the ABC-TV *Insiders* studio on Sunday, 29 April, I did not "admit" to anything. He merely asked me whether I had seen *The Death Series* in full. I replied in the negative. I added that I had viewed the extract on the Channel 9 News and that I had watched the Channel 4 *Dispatches* program, which showed footage from the video, when I was in London last January.

This should have come as no surprise to Marr. After all, in my *Sydney Morning Herald/West Australian* column, published on 17 April, I did not claim to have viewed the entire DVD. Rather, I made specific reference to the Channel 9 News report and to the *Dispatches* program. It is disingenuous for David Marr to imply otherwise.

I guess we just disagree on this issue. Marr wants to defend the right of an Islamist sheik to declare that Muslims "want to have children and offer them as soldiers defending Islam" and to advocate (while imitating a snorting pig) that someone should "come and kill ... the Jew" – and for such material to receive a PG rating from the Office of Film and Literature Classification. I hold a different view. I do not believe that free speech entails that individuals should be absolutely free to publicly advocate the murder of Jews – or of Christians or Hindus or Muslims or gays or heterosexuals or whomever.

In any event, I invited David Marr to discuss the views presented in *His Master's Voice* at the Sydney Institute. Alas, he never got back to me.

Gerard Henderson

Andrew Bolt

David,

I'm intrigued. Did you deliberately misrepresent me on page 51 of your *Quarterly Essay*, or were you merely careless – once more led astray by the melodramatic stereotypes which haunt your imagination?

To remind you:

You pick me out as one of those who have helped to corrupt public debate on the grounds that I've listed artists who have said truly preposterous things that even you cannot defend, and furthermore I "want to see these artists punished for their anti-democratic tendencies."

Really? I do?

True, I want them to suffer the scorn that anti-democratic tendencies should rightly attract in free society.

But actively punished by the state in an "echo of the rhetoric Moscow used"? What evidence do you have for that ludicrous proposition?

Ah! Here is your proof:

"[Bolt] argues it's high time Canberra stopped 'the liberal use of taxpayers' money … to fund attacks on our mainstream culture or institutions.'"

I don't want to get too subtle on you, but you've pulled a five-year-old quote completely out of context to misrepresent what I've actually argued quite clearly for years.

The quote was in fact part of an observation that the way we fund artists now – often directly, and through bureaucracies, not audiences – encourages a group-think that is deadening, and an indifference for audiences that has made our arts seem largely irrelevant and utterly marginal. The utter failure of our film industry and the slim sales of our novelists are two signs of this withering.

My solution is not at all to "punish" artists for "being out of line," or even to stop "the liberal use of taxpayers' money," and I have never suggested any such thing as you claim.

It is – as I've even told you personally – to subsidise audiences, not artists directly, to better encourage the primary connection on which great art thrives.

If the audiences so subsidised clamour for "attacks on our mainstream culture and institutions," then so be it. I want no punishment of the artists who produce such attacks, any more than I'd want to be punished for my own.

But if these subsidised audiences decide instead in droves to see something that speaks deeply to them, from a warm but not uncritical understanding of their home and time, so much the better.

How strange that such proposals – for increased funding, for more diversity, for a broader debate and for art that commands greater audiences and attention – should instead be misrepresented so egregiously as an attempt to stifle what it actually hopes to encourage.

I do hope the rest of your essay is less deceptive. Less inclined to paint stereotypical villains.

But I fear the worst, and reading you only confirms it.

Andrew Bolt

HIS MASTER'S VOICE	Response to Correspondence and Epilogue

David Marr

The Attorney-General and I knock along as best we can. In that spirit, I want to assure him I also think Australia should be kept safe from terrorism, that our way of life should be protected, that Australians should not be derided as unthoughtful or unsophisticated and that this is a wonderful country. When the national anthem plays, we'll both be on our feet singing.

But there's no glossing over the differences between us. Whistleblower protections? There are none for public servants who want to go public. Shield laws for journalists? There are none where Commonwealth public servants leak to the press. Sedition laws? Journalists aren't the only ones in trouble while Ruddock refuses to provide safeguards recommended by the Australian Law Reform Commission to protect writers, satirists, academics, film-makers, painters and trade unionists. Habeas corpus can safely be suspended to question terrorist suspects? Not unless he learns the lessons of the Haneef debacle. Voters have the final say? Of course – but this government has gone to unprecedented lengths to keep them from knowing what they need to know before they reach the ballot box. That's what it's all about.

Peter Shergold's response is thoughtful and thought-provoking. But his assurance that Canberra public servants don't need to fear fundamental moral challenges in their work raises the ghost of the *Tampa* crisis and the efforts of his ruthless predecessor, Max Moore-Wilton. The moral challenges were there all right, but the bureaucrats squibbed them. Marian Wilkinson and I wrote in *Dark Victory*:

> The Canberra bureaucracy was in shock ... Some of them won-
> dered if, somewhere ahead, there would be lines they were unwill-
> ing to cross. The military found those lines. The civilians never did.
> The work was distasteful but they were caught up in the same

atmosphere of crisis that dazzled politicians and the press. The bureaucrats felt professionally obliged to deliver what the Prime Minister and the government wanted; not frank and fearless advice but solutions to "the predicament Rinnan has put us in". They were expected to "leave their baggage" – their personal values – "at the door". They did.

No government in Canberra need be nervous of a bureaucracy that delivered the *Tampa* operation without an open revolt. We're talking obedience here. Even loyalty. During that ten-week crisis there were not even any great leaks to the press – partly, perhaps, because senior bureaucrats took the precaution of putting as little as possible in writing.

What Shergold doesn't address here is the extraordinary level of secrecy he and Moore-Wilton have demanded from the public service. Journalists coming home from stints in Washington are struck afresh by Canberra's hostility to the free flow of information from bureaucrats: not secrets, just useful stuff about how the government is going, the sort of information senior Washington public servants give over the phone to any hack who rings. Why not in Canberra? Why have the rules been tightened under Howard? Shergold doesn't say.

I fell in love with the concept of *reductio ad absurdum* when I was a first-year university student. I'm still devoted to the idea that ridiculous outcomes point to ridiculous assumptions hidden somewhere deep inside arguments. Shergold's eloquent exposition of the policing that "good governance" requires led to the prosecution of Allan Kessing. Everyone who catches a plane in this country is safer today because Kessing's neglected report into airport security turned up on page one of the *Australian*. Millions were thrown at problems he uncovered. But in the interests of good governance, Kessing was convicted under the *Commonwealth Crimes Act* and escaped jail by a whisker.

Perhaps in honoured retirement after serving a term or so in an embassy abroad, Shergold will give us his answer to the deeper question he skirts here: when do the needs of the public override the needs of government? I suspect his answer is: almost never.

News Limited finds itself in a tricky position. Its boss, John Hartigan, is the driving force behind a coalition of media proprietors and journalists urgently lobbying Canberra "to redress the erosion of free speech in this country." But the dozens of papers he runs are still backing John Howard. So Harto is simultaneously advancing the Right to Know campaign while going easy on the Prime

Minister. He writes here: "It is, quite frankly, unhelpful to lay all the blame at the gates of the Lodge."

He lays hardly any at all – and accuses me of paranoia, dogged ideology and jaundiced dislike of the man for seeing Howard's hand behind the problems we're facing. How can we not? Howard has been putting his stamp on this country for the last eleven years. That cultural transformation is ceaselessly celebrated in the Murdoch press. It's true the problems pre-date Howard and cross party lines – which I acknowledge in the essay – but we can't come to grips with them unless we examine the role of this supremely effective politician.

News Corp's riding instructions were revealed with disarming frankness by the editor of the *Weekend Australian*, Nick Cater, when defending the *Oz*'s editorial attacking me, Clive Hamilton and Robert Manne as "psychotic" lefties after *His Master's Voice* appeared in June. Cater told *Crikey*:

> The *Australian* and News Corp are at the forefront of the campaigning to ensure the right to know. This is very distinct and different from the claims made by books like *Silencing Dissent* ... in which they claim there is effort by the Howard government to suppress political debate and silence voices of the left. As we made clear in our editorial, we find no evidence that there is any politically motivated attempt to silence dissent in Australia.

Analysis like this doesn't fill me with high hopes for the Right to Know campaign. What's happened to public debate in Howard's Australia hasn't happened by accident.

Despite what Tom Switzer says, I don't claim debate has been silenced. I don't blame Howard for virtually every act of dishonesty and suppression in today's Australia. I don't claim Australians are philistine and stunted. Never have. I say as clearly as the language allows that Patrick White's Australia is not the Australia of today. I have never defended violent demonstrations. I leave to others the language of autocracy and authoritarianism. There was never a time I believed my political friends and I set the nation's agenda. I wish. Nor do I claim to represent Australia's conscience: I can only speak for my own.

Over the years it's puzzled me why, when there are so many real differences between us, Switzer invents more. But heroic struggles with straw men are so often how it goes on the *Australian*'s opinion pages which he has edited since 2001. His latest – not uncomplimentary – response to my essay comes after a bollocking from him when it first appeared in June, a damning review – no complaints from me there – and that florid, full-drop editorial naming me among the leaders of the "psychotic left". It has to be said, the *Oz* gave *His Master's Voice* a wild welcome.

I wasn't thinking of it then, but now I see that any essay attacking Howard's style of debate had to tangle with the *Australian's* too. They are soul brothers in controversy. It is not by accident, I think, that almost the only thing the leader, the reviewer and now Switzer actually quote from the essay is my argument that fifty years down the track Patrick White's "exaltation of the average" is back in a big way. I went on to say:

> A great deal of effort by both the government and its outriders in the press has been devoted to mooring Australian public values in this thing they call the mainstream. Abuse of those who challenge mainstream ideas is routine in Howard's Australia.

And routine in the pages of the *Oz*. Switzer is at it again here – though politely – when he claims "the great weakness" of my argument is that I detest a number of ideas that have solid democratic backing. He's right: I do. But so what? Isn't this how ideas shift and change in a democracy? Isn't this how the "conservative" values he touts triumphed over the old "liberal" consensus? Do I have to shut up because the people have decided otherwise? And if the main-stream now has the democratic right to sweep all before it these days, how come Switzer applauds Howard masterminding a political operation to ban euthanasia against the wishes of something like 80 per cent of the Australian community? He doesn't say.

Howard and the *Oz* only cite the mainstream when it suits. What all thought-ful citizens know – and I don't exclude Switzer – is that good policy is not always popular. And vice versa. Disentangling what's wise from what's popular has been a core task of public debate from the beginning of time. We're all in the same game here: it's called democracy. And in Howard's Australia we all face the same problem: "The degree to which the flow of information that generates or fuels informed debate has been stifled." The words are those of Switzer's boss, John Hartigan.

Yes, Howard is for the most part polite. He doesn't hit the phones as Keating did and Rudd does. He doesn't sue. He doesn't fight in the gutter. But his minis-ters do and they do so with his blessing. Tony Abbott, Bill Heffernan and Eric Abetz are grubby operators in debate, and on promotion to the portfolio of Workplace Relations, Joe Hockey quickly earned a reputation as a thug. That Switzer defends the tactics he employed against David Peetz, I find deeply troubling.

A minister testing a critic's "perceived biases and connections" is one thing, but what about the point Switzer fails to mention: Hockey attacking Peetz as a terrorist sympathiser? I emailed to check he hadn't accidentally overlooked this. He hadn't. Switzer told me he rated Hockey's claim a cheap shot that did the government no credit, but nothing more: "I just didn't want to get bogged down in more of your examples/disputes that could easily be open to interpretation."

Interpretation? No honest commentator could conclude, on the evidence of the professor's clumsy poem, that Peetz endorsed terrorism. This was a lie deliberately deployed by Hockey and Abetz to ruin the reputation of an effective critic of WorkChoices. That a couple of ruthless politicians would throw such muck is par for the course in Howard's Australia. So, alas, is the failure of a leading newspaper's opinion editor to name it for what it is. Easily open to interpretation? Bullshit. The *Oz* and the government see eye to eye on Peetz and eye to eye on the rough-house rules of debate that have flourished in the Howard years. If it does the job, any cheap shot will do. Even a lie.

When the row over Sheik Feiz broke in January this year, the *Sydney Morning Herald* assigned a team of journalists to plough through his DVDs and report what Feiz was actually preaching. That's not Gerard Henderson's way. He condemned them sight unseen. Had he confessed he'd watched only a couple of minutes out of a couple of hours of these sermons, I wouldn't have been so perturbed by his column. But Henderson broke two rules that day: he didn't come clean and he didn't watch for himself what he was demanding be banned.

Like the Office of Film and Literature Classification, I see banning as an absolute last resort. I'm not sure Henderson does. Nor am I certain he would change his mind if he actually watched the DVDs. Views can differ. To my mind they're confronting but not dangerous. In particular, I don't believe Feiz is telling contemporary Muslims to go out and kill Jews. Henderson may come to another conclusion altogether, but surely he must first watch Feiz at work. And so should anyone who cites these unseen sermons as a reason to send police into religious bookshops. Where would it end? The rhetoric is the Sheik's, but the teachings are the Prophet's. And Islam is not the only faith that preaches slaughter of unbelievers at the end of time.

Gerard, here is what happens in *Signs of the Hour* as Isa (Jesus) prepares for a final showdown with the "one-eyed monster, the biggest liar, the devil" who is roaming the earth performing false miracles. Most of his followers are Jews, the Sheik informs his unseen audience. But the anti-Christ has also recruited "ladies" and "ignorant Arabs" who are all about to meet their fate as Isa descends from heaven supported by angels and scattering beads of sweat that turn into pearls. Unbelievers, we are told, drop dead merely by a glimpse of His face or a whiff of His breath.

Inside the "Eastern Gate of Lead" somewhere in the Middle East, Isa takes His stand at the head of a Muslim army. He orders the gates opened:

> Behind the gate will be none other than that one-eyed liar with an army of pigs, an army of Jews. And they will be bearing – all of them – a sword and a shield. They will be [the sheik makes ugly grunting noises] all of them. You can hear the noise.

At the sight of Isa, the anti-Christ:

> starts dissolving like salt dissolves in water and he run away but Isa will call out to him, you will not die until I strike you with my sword … and all the Jews, his army, will be defeated. There will be nowhere to hide, not a tree, not a stone, not a creature except this inanimate or animate creature will say, Oh Muslim behind me is a Jew come and kill him. Oh Muslim behind me is a Jew come and kill him. Thus Isa and the companions will win over these evil people.

Isa then turns his attention to the rampaging monsters Gog and Magog:

> He will send an insect to destroy Gog and Magog by biting them on their necks and next day they will all perish as one … and He will send birds like the camels' necks, big birds, who will carry Gog and Magog and take them wherever Allah wills and that will be the end of them. After that peace and security will prevail on earth, whereby the lion will breed with the camels, the tigers with the cattle …

Andrew Bolt's message hasn't changed. He reckons artists who belittle "the mainstream" or "Western society" should not be given public money and he's been saying so for years. Here is Bolt in 2002 attacking the choice, a decade earlier, of Mandawuy Yunupingu as Australian of the Year:

This is how the Tribe maintains itself – this, and the liberal use of taxpayers' money by the ABC, our universities, our research grant system and the Australia Council to fund attacks on our mainstream culture or institutions … Ah, the Tribe. Closed minded and closed off. Still, if it had our ear as well as our cash, it would be dangerous.

Here he is last November making the same complaint about artists who don't represent "us":

> Peter Carey called our PM a toadie of an America led by a coward, but that was just a whiff of the full tin of cat's meat. The weird news this past month tells us enough is enough. These people, so many shrouded in black and living in enclaves like Brunswick, must now become assimilated. What a disaster it's been to instead pay them special grants allowing them to keep aloof from the rest of us.

After attacking "prize-winning and taxpayer-funded" Phillip Noyce, Stephen Sewell, Andrew McGahan, Richard Flanagan and Marion Halligan, Bolt declares:

> the urgent challenge is to help this alienated minority feel welcome in this country, and to reward them for joining in, not staying apart. Assimilation is the key. Their ghettos of hate must go. How we'd love to call them ours.

This is trademark Bolt. He's been demanding these people be stripped of grants and prizes for years. It's punishment for politics. Give me a tin of stewed kangaroo any day: what I get when I open this can is a strong whiff of the old GDR.

Epilogue

It's all still going on. As the elections approach, the government's behaviour grows worse and the Opposition's response more supine. Even little victories along the way turned sour.

Attacks on David Peetz tailed off as the government realised silencing a handful of academic critics wasn't going to save WorkChoices. The name was abandoned, fresh tests introduced and tens of millions of dollars began to pour into a campaign to sell the latest version of WorkChoices and – if possible – save the government. The last of these tactics raises an issue I didn't cover in the essay: the corruption of public debate through tax-funded political advertising. Hawke and Keating led the way here, but Howard is rorting the public purse on an unprecedented scale.

Woolgrowers, having spent about $10 million, abandoned their hopeless case against the anti-mulesing forces of People for the Ethical Treatment of Animals (PETA). But on 15 August Peter Costello brought into parliament his long-promised legislation to allow the Australian Competition and Consumer Commission to prosecute those who call for customer boycotts – of mulesed wool, or rugs woven by slave labour, or soft drinks that make kids obese. These aren't Costello's examples, of course. He told parliament: "The government reaffirms its commitment to stand up for small business against thuggery and intimidation. It is vital, both for our economy and our way of life."

The Tranny Cops arrested after Dick Cheney's visit to Sydney were acquitted. In a scathing decision that drew on the Village People, Popeye and *The Pirates of Penzance*, Sydney magistrate David Heilpern pointed out to senior police from the APEC squad that Sarah Harrison and Anika Vinson – wearing ink moustaches and dressed in dark blue overalls embroidered with the motto "Cop it Sweet!" – were joking. Police remain unamused. Elaborate plans have been unveiled to lock Sydney down for APEC. More is at stake than keeping the world's leaders safe. The NSW Police Minister, David Campbell, has declared demonstrators will be kept well away from the summit to avoid any "embarrassment" to VIPs.

Allan Kessing was eventually given a nine-month suspended sentence for the crime of leaking reports of the woeful security at Sydney airport – leaks that led to a revolution in arrangements at all major airports in this country. NSW District Court judge James Bennett found no fault whatever in the Customs Department for failing to act on Kessing's reports for nearly two years. "Whether or not it is appropriate to view the offender in the heroic light with which he has been bathed by some," concluded the judge, "there was no justification whatsoever for communicating the contents of these reports." Kessing plans to appeal.

Ruddock struck out again with the state and territory attorneys-general in Hobart in late July when they refused to endorse his plan to ban all material "advocating" terrorism. A discussion paper outlining the proposals had drawn a barrage of objections. Even the convenor of the government's Classification Review Board, Maureen Shelley, observed the plan was "a significant departure from current practice" and would see the banning of just about any text praising any act of terrorism. Rebuffed by the Ags, Ruddock announced he would fly solo and impose the new rules from Canberra. Rudd assured him of Labor's support.

But these are sideshows compared to the shame of the days in which I'm writing – the few days in which elaborate plans to transform the governance of Aboriginal Australia are being guillotined through parliament. Over five hundred pages of legislation were published one day and passed by the House of Representatives the next. "This process is corrupting this parliament," declared Greens senator Bob Brown. Labor was more or less mute. A week later, the bills were law and the lives of 40,000 Australians – black Australians – living in the Northern Territory will change profoundly.

Howard has form on this: the security legislation of 2005 kept secret for as long as possible then rammed through parliament; the hugely complex Work-Choices rushed through both chambers in 2006 without the scrutiny that might have saved the government from today's backlash; and now the radical Emergency Response in the Northern Territory hurried through both houses. These are nation-changing measures. Why the rush? Howard is assured of them passing. He has the numbers. So why curtail deliberation in ways unimaginable in Britain or the United States?

Those who acknowledge the corruption of public debate in Australia but somehow absolve the Prime Minister of responsibility must accept these tactics were devised or approved by him. Though he cites a "strong" parliament as one of the bulwarks of democracy, he is happy to turn the legislative process into this sort of charade. That the problems won't disappear when he leaves office is no excuse. We're dealing with here and now. Nor can the looming elections be used as some kind of rough political justification for these tactics. No. This is John Howard. It's what he thinks of democratic debate. It's him.

David Marr

Waleed Aly is a lecturer in the school of Political and Social Inquiry at Monash University, where he also works in the university's Global Terrorism Research Centre. He is the author of the just-published *People Like Us: How Arrogance is Dividing Islam and the West*.

Andrew Bolt is a columnist for Melbourne's *Herald Sun*.

Julian Burnside QC is the president of Liberty Victoria.

John Hartigan is the chairman and chief executive of News Ltd in Australia and a founding member of the Right To Know Coalition.

Gerard Henderson is the director of the Sydney Institute and a columnist for the *Sydney Morning Herald*.

Ian Lowe is emeritus professor of science, technology and society at Griffith University and president of the Australian Conservation Foundation. He studied engineering and science at the University of New South Wales and earned his doctorate in physics from the University of York. In 1991 he gave the ABC's Boyer Lectures. He is the author of many books, including *A Big Fix* and *Living in the Hothouse*.

David Marr is the multi-award-winning author of *Patrick White: a Life* and *The High Price of Heaven*, and co-author with Marian Wilkinson of *Dark Victory*. In a career spanning over thirty years, he has written for the *Bulletin* and the *Sydney Morning Herald*, been editor of the *National Times*, a reporter for *Four Corners* and, until recently, presenter of ABC-TV's *Media Watch*.

Philip Ruddock is the Commonwealth attorney-general and a former minister for immigration and for multicultural and indigenous affairs.

Peter Shergold is the secretary of the Department of Prime Minister and Cabinet.

Joan Staples is a visiting fellow at the law faculty of the University of New South Wales, researching the role of civil society in the democratic process. She has worked for many NGOs, including the Cape York Land Council, the Australian Conservation Foundation and the Diplomacy Training Program, affiliated with the University of New South Wales.

Tom Switzer has been opinion-page editor of the *Australian* since 2001. He is a regular contributor to *Quadrant* magazine and the *Wall Street Journal* editorial page.

Subscribe to Quarterly Essay

POST OR FAX THIS FORM TO: Quarterly Essay, Reply Paid 79448, Melbourne VIC 3000
Freecall: 1800 077 514 Fax: 61 3 9654 2290 Email: subscribe@blackincbooks.com

SUBSCRIPTIONS Receive a discount and never miss an issue. Mailed direct to your door.

1 year subscription (4 issues): $49 a year within Australia incl. GST. Outside Australia $79.

2 year subscription (8 issues): $95 a year within Australia incl. GST. Outside Australia $155.

* All prices include postage and handling.

BACK ISSUES Please add $2.50 postage and handling to your order (or $8.00 for overseas orders).

☐ **Issue 1** ($9.95) Robert Manne *In Denial: The Stolen Generations and the Right*
☐ **Issue 2** ($9.95) John Birmingham *Appeasing Jakarta: Australia's Complicity in the East Timor Tragedy*
☐ **Issue 4** ($9.95) Don Watson *Rabbit Syndrome: Australia and America*
☐ **Issue 5** ($11.95) Mungo MacCallum *Girt by Sea: Australia, the Refugees and the Politics of Fear*
☐ **Issue 6** ($11.95) John Button *Beyond Belief: What Future for Labor?*
☐ **Issue 7** ($11.95) John Martinkus *Paradise Betrayed: West Papua's Struggle for Independence*
☐ **Issue 8** ($11.95) Amanda Lohrey *Groundswell: The Rise of the Greens*
☐ **Issue 10** ($12.95) Gideon Haigh *Bad Company: The Cult of the CEO*
☐ **Issue 11** ($12.95) Germaine Greer *Whitefella Jump Up: The Shortest Way to Nationhood*
☐ **Issue 12** ($12.95) David Malouf *Made in England: Australia's British Inheritance*
☐ **Issue 13** ($12.95) Robert Manne with David Corlett *Sending Them Home: Refugees and the New Politics of Indifference*
☐ **Issue 14** ($13.95) Paul McGeough *Mission Impossible: The Sheikhs, the US and the Future of Iraq*
☐ **Issue 15** ($13.95) Margaret Simons *Latham's World: The New Politics of the Outsiders*
☐ **Issue 16** ($13.95) Raimond Gaita *Breach of Trust: Truth, Morality and Politics*
☐ **Issue 17** ($13.95) John Hirst *'Kangaroo Court': Family Law in Australia*
☐ **Issue 18** ($13.95) Gail Bell *The Worried Well: The Depression Epidemic and the Medicalisation of Our Sorrows*
☐ **Issue 19** ($14.95) Judith Brett *Relaxed and Comfortable: The Liberal Party's Australia*
☐ **Issue 20** ($14.95) John Birmingham *A Time for War: Australia as a Military Power*
☐ **Issue 21** ($14.95) Clive Hamilton *What's Left?: The Death of Social Democracy*
☐ **Issue 22** ($14.95) Amanda Lohrey *Voting for Jesus: Christianity and Politics in Australia*
☐ **Issue 23** ($14.95) Inga Clendinnen *The History Question: Who Owns the Past?*
☐ **Issue 24** ($14.95) Robyn Davidson *No Fixed Address: Nomads and the Fate of the Planet*
☐ **Issue 25** ($14.95) Peter Hartcher *Bipolar Nation: How to Win the 2007 Election*
☐ **Issue 26** ($14.95) David Marr *His Master's Voice: The Corruption of Public Debate Under Howard*

PAYMENT DETAILS I enclose a cheque/money order made out to Schwartz Publishing Pty Ltd. Please debit my credit card (Mastercard, Visa or Bankcard accepted).

Card No. ☐☐☐☐ ☐☐☐☐ ☐☐☐☐ ☐☐☐☐ ☐☐☐☐

Expiry date   /   Amount $

Cardholder's name   Signature

Name

Address

Email

Subscribe online at www.quarterlyessay.com

www.ingramcontent.com/pod-product-compliance
Lightning Source LLC
Chambersburg PA
CBHW040154200326
41519CB00044B/7604